81 Reasons That Defy the Age of the Earth

Is the Earth Really 4.6 Billion Years Old?

Author: Rafael E. Dickson Castillo

Table of Contents

INTRODUCTION *12*
Author's Motivation 12
Overview 12
Critique of the Scientific Consensus 13
Book Organization 13
Expectations for the Reader 13
Evolution of Book Versions 14
The Challenge of Scientific Consensus 14

ON DATING METHODS. *22*

Is Dating a Fallible Science? Radioisotopes, Redshift, and Background Radiation **22**

1) **Radioactive radio isotopes:** 25

2) **Redshift (RedShift) and Microwave Background Radiation (CMB)** 60

PART I: Astronomical and Cosmic Evidence 89

3) **The shrinking of the Sun:** 89
Implications for Evolutionary Theory: 89

4) **Cosmic dust on the Moon:** 91
Implications for Evolutionary Theory: 91

5) **The Moon's Retreat:** 92
Calculating the distance in the past: 92
Impact on tides: 93
Consequences for continents and life: 93

6) Cooling of Jupiter and Saturn: 94
Evidence of Cooling: 94
The Case of Io, a Moon of Jupiter: 95
Implications to the Chronology of the Solar System: 95

7) Sirius Transformation: 96
Historical and scientific evidence: 96
Possible explanations: 97
Implications for evolutionary times: 97

8) Evidence of cosmic radiation: 98
Evidence and analysis: 99
Analogy: 99
Implications for the age of the universe: 99
Aspects that are questioned in the standard cosmological chronology: 101

9) Accelerated Stellar Evolution: 102
Observations of Rapid Transformations: 102
Examples of Rapidly Evolving Stars: 103
Implications for Stellar Evolutionary Times: 103

10) Recent supernovae: 104
Evidence of the scarcity of supernova remnants: 104
Implications for the age of the universe: 105

11) Interplanetary dust: 106
Evidence observed: 106
Dust accumulation estimates: 106
Implications for the age of the solar system: 106

Common counterarguments: 107

12) Alignment of the planets: 108
Evidence observed: 108
Orbital stability: 109
Questioned models: 109
Implications for the age of the solar system: 109

Part II: Geological Evidence 111

13) The erosion of the continents: 111
Evidence observed: 111
Implications for the age of the Earth: 112
Common counterarguments: 112

14) Corrosion in Niagara Falls: ... 113
Evidence observed: 113
Implications for the age of the Earth: 114
Common counterarguments: 114

15) The Mississippi River Delta: . 115
Evidence observed in the Mississippi Delta: 115
Comparison with other deltas: 115
Implications for the age of the Earth: 116
Common counterarguments: 116

16) Formation of stalactites and stalagmites: 117
Evidence observed: 117
Implications for the age of the Earth: 119
Common counterarguments: 119

- 17) **The oldest coral reef:** **120**
  - Evidence observed: 120
  - Implications for the age of the Earth: 121
  - Common counterarguments: 122

- 18) **Marine sediments:** **122**
  - Common counterarguments: 124

- 19) **Pressure in oil tanks:** **125**
  - Evidence observed: 125
  - Implications for the age of the Earth: 126
  - Common counterarguments: 127

- 20) **Evidence in Tectonic Plates:.** **127**
  - Evidence observed: 127
  - Implications for the age of the Earth: 129
  - Common counterarguments: 129

- 21) **Erosion of volcanoes:** **130**
  - Evidence observed: 130
  - Implications for the age of the Earth: 131
  - Common counterarguments: 132

- 22) **Salinity of the oceans:** **132**
  - Evidence observed: 132
  - Implications for the age of the Earth: 134
  - Common counterarguments: 134

- 23) **The Rocky Mountains:** **135**
  - Evidence observed: 135
  - Rocky Mountain Erosion Rate: 136
  - Implications: 137
  - Common counterarguments: 139

- 24) **Fresh Sedimentary Rocks:** ... **139**

Evidence observed: 139
Implications for the age of the Earth:
.... 141
Comparison with evolutionary
models:... 141
Common counterarguments:......... 141

25) Expansion of the Sahara:...... 142
Evidence observed: 142
Implications for the age of the Earth:
.... 144
Comparison with evolutionary
models:... 144
Common counterarguments:......... 145

26) Colorado Canyon: 145

27) Basalt Pillars:...................... 147
Notable examples of basalt pillars
include:... 147

Part III: Biological Evidence............... 149

28) Genetic Mutations: 150
Explanation with apples: 150
Relationship to evolution: 151
Another simple example: 151
Counterargument (from the
evolutionary perspective):...................... 152
Errors in evolutionary reasoning: ... 154

29) Extinction of Species: 157

**30) Biological Clocks
Dendrochronology 158**

31) Soft Tissue Fossils:............... 163

32) Ancient Bacteria:.................. 165

33) Mitochondrial DNA Analysis: 167

34) Ecosystem Collapse: 168

35) Animal Survival Record: 170

Part IV: Physical and Atmospheric Evidence *172*

36) Atmospheric Helium and Young Earth 172

37) Terrestrial Helium: A Mystery 173

38) Helium Leakage and Radioactivity in Rocks 173

39) The Evidence for Helium in Polar Ice 174

40) The Powerful Force of Helium in the Universe 174

41) Amount of nitrogen in fossils: 175

42) Oxygen in the atmosphere: .. 177

43) The Ice Age at the Poles: 178

44) The balance of CO2 in the atmosphere. 179

45) Decreased Magnetic Field.... 181

46) Weathering of Oceanic Crust 183

Part IV: Physical and Atmospheric Evidence *186*

47) Comet Analysis: 186

48) Stellar Laughter.................... 187

49) Geothermal Balance and Earth Cooling. 189

50) "Fossilized Trees in Multiple Strata" 191

51) The Magnetism of Rocks. 193

52) "Meteorite Shortage in Ancient Layers" 197

53) "Inconsistencies in Isotope Dating in Volcanic Rocks" 200

54) Polonium Halos in Rocks 201

Part V: Biological, Geological and Population Evidence. 204

55) Oldest historical records: 204

56) Accelerated evolution in fossils: 207

57) Genetic material in ancient fossils: 209

58) Fossil remains without evolutionary change: 211

59) Limited genetic diversity: 212

60) Distribution of ancient languages: 213

61) Lack of ancient human remains: 214

62) Distribution of civilizations: . 215

63) "Potassium Isotopes and Volcanic Rocks" 216

64) Cambrian Explosion Theory:. 217

65) Fossils in suffocation position:219

66) Human Presence in Coal Layers. 220

67) Recent Origins of Agriculture 220

68) Mutations and Natural Selection 222

69) Fossils and Rapid Formation. 223

70) On the same amber. 224

71) Marine Fossils in the Mountains. 225

Part VI: Chemical Evidence *227*

72) Anomalies in Radioactive Decay of Uranium 227

73) Radioactive Decay Abnormalities of Thorium 228

74) Rapid sedimentation in fossils: 229

75) Recent Supernovae: 231

76) Accelerated Stellar Evolution. 231

77) Venus and its Mountains. 232

78) Balance of the Solar System. 233

79) Expansion Speed of the Universe. 235

80) CFCs in Polar Ice 236

81) Changes in Sea Level and Coastal Strata 237

Conclusion *239*

Explanatory Note on Sources: *245*

INTRODUCTION

Author's Motivation

This work is born from a deep commitment, cultivated over 30 years, to investigate and understand evidence that challenges the paradigm of evolutionary time and underpins a recent chronology for our planet and the universe.

In this path of study and reflection, I have found arguments – some even through serendipity – that I consider essential to understand our history from a creationist perspective.

It is not my intention to defend God, as His eternal power and deity are made visible through created things.

My goal is, rather, to expose and share the journey that has led me to these convictions, in the hope that this evidence will serve as a light and encourage reflection in those who seek a deeper understanding of reality.

Overview

The intention of this work is to present scientific evidence that, from different fields – cosmology, geology, biology and others – offers a coherent testimony in support of a young Earth.

My goal is not to prove every piece of evidence.

I don't have to prove, for example, that the Moon is moving away from Earth at a rate of nearly 4 inches per year. I have not said it. The "scientists" of the "scientific community" say so.

However, I do ask myself the question of where it must have been a million or even 66 million years ago, and what the implications of that distance might have been for our planet. Similarly, I raise similar questions about other evidence, and when there are arguments to the contrary, I also present and analyze them.

This body of evidence is presented not only as a refutation of evolutionary time, but also as a solid alternative that aligns with the biblical account and belief in a Creator.

To readers committed to discovering truth and willing to examine the basis of their knowledge, this book invites them to reconsider the accepted chronology and opens a window into a perspective that has been ignored or minimized by many.

Critique of the Scientific Consensus

In an age subjugated by what I call an "information dictatorship," we find that many scientific voices and perspectives have been relegated or even silenced.

Here, it is not a question of rejecting science, but of examining how dominant assumptions, which are usually accepted without question, shape a seemingly indisputable consensus.

This book positions itself as a resource to challenge the "*status quo*," encouraging an authentic and dispassionate search for truth.

Science advances by questioning, and this project seeks precisely that: to open space for informed and supported questioning, rather than blind acceptance.

Book Organization

To facilitate the study of each evidence, the content is organized into seven sections, inspired by the biblical symbolism of the number seven, which represents fullness and divine perfection.

These sections cover evidence in astronomical, geological, biological, and other fields. Through this structure, the reader will be able to see how each element connects to a complete vision of the history of the Earth and the universe, contrasting with the "six days" of creation and the multiple theories that have tried to explain existence from an exclusively materialist perspective.

Expectations for the Reader

This book invites you on a journey of questioning and discovery. Each piece of evidence is presented as a tool for deep thinking and with an open mind, in an intellectual exercise that challenges the reader to consider another side of the story.

Throughout these pages, the expectation is not to impose a belief, but to inspire a genuine interest in examining, comparing, and discerning what has been learned. What conclusions will this evidence bring to those who contemplate it? Only the reader has the key to unlock that answer.

Evolution of Book Versions

The first version of this book was born unexpectedly. It was not a project conceived from the beginning for these purposes; rather, it emerged as a collection of data and information that I gathered and saved on my laptop over the years.

This initial compendium, of simple language and without the weight of unnecessary references or argumentative extensions, was mainly designed to be listened to on devices such as Kindle, which makes it less technical and more accessible.

As a Christian, I find satisfaction in that first version, as it conveys the essential foundations of the evidence supporting a young Earth and a recent universe.

However, recognizing that some people are looking for a more detailed and technical presentation, I responded to the criticism and interest of a more demanding audience by developing a second version, which was ultimately the one that was translated into English.

This second edition improves the arguments, introducing studies and sources where possible and removing the emotional burden that, as a Christian, I might add. However, the concept and purpose remain unchanged: *to present a solid alternative that invites questioning of the accepted standard chronology.*

The Challenge of Scientific Consensus

One of the biggest challenges facing this work is that much of the ideas and evidence presented here come from scientists and institutions that are not considered part of the "scientific consensus" of the mainstream community. This scientific consensus, dominated by theories such as the Big Bang, stellar evolution, and the expansion of the universe, rests on the idea of a cosmos developing over billions of years.

From the perspective of many mainstream scientists, proponents of creationist models or alternative cosmologies fall outside the mainstream of science, as their interpretations are often based on religious or philosophical convictions that differ from the current naturalistic framework (that's what they say). Therefore, it is common for those who defend the standard or official model to discredit or ignore these approaches, arguing that they do not comply with the conventional scientific method.

However, this book aspires to give voice to these alternative interpretations, not to impose beliefs, but to open an informed dialogue about the possibility of a younger universe, presenting foundations based on observation and genuine questioning.

This effort responds to the conviction that the search for truth must be broad and allow a space for critical research. Ultimately, my intention with this second version is not to turn a deaf ear, but to offer readers the option of taking an open look at these arguments and drawing their own conclusions.

This book demonstrates that the models and theories proposed by scientists from institutions such as AiG *(Answers in Genesis)* or the ICR *(Institute for Creation Research)*, although not widely accepted, apply scientific principles in their analyses.

These approaches draw on astronomical observation, physics, and thermodynamics to raise their objections, without relying on the Bible or philosophies. The debate lies in the interpretation of these data and in the assumptions that underpin each model.

For those who hold a creationist worldview or question strict naturalism, these scientists and organizations represent a valid contribution to scientific discussion, even if they are marginalized by the dominant consensus. This "Scientific Community" – wrongly called that when it becomes exclusive – should not dismiss alternative voices.

The estimate that the Earth is about 4.6 billion years old is the result of two centuries of scientific research, and has become a widely accepted and official calculation, based primarily on radiometric dating methods, which analyze the decay of radioactive isotopes in Earth's rocks and minerals. But how reliable are these methods? Can we really say with certainty that the Earth is billions of years old, or is there evidence to suggest a much more recent chronology?

Several men of science played a crucial role in the development of methods for estimating the age of the Earth. One of the first to try was Lord Kelvin (1824-1907), who calculated the age of the planet based on its thermal cooling. However, due to ignorance of radioactivity at the time, Kelvin significantly underestimated this age.

In the late nineteenth century, Marie and Pierre Curie's discoveries of radioactivity laid the foundations for radioactive dating, a field later expanded by Henri Becquerel.

In 1904, Ernest Rutherford introduced the idea of using radioactive decay as a natural "clock" to measure the age of rocks, paving the way for what we know today as radiometric dating.

Finally, in 1956, Clair Cameron Patterson calculated the age of the Earth at 4.55 billion years using meteorite samples and the uranium-lead dating method. Although he was not awarded the Nobel Prize, his contribution was instrumental in establishing the age of the Earth that is currently accepted.

However, the use of meteorite samples and the uranium-lead method are today in question, in light of the numerous evidence against them, also calling into question the validity of this acceptance.

Over the years, some scientists have ignored or overlooked evidence that could suggest a lesser Earth.

Radioactive dating methods are considered revolutionary and generally accurate; however, they are based on several questionable assumptions, such as the constancy of radioactive decay rates and the absence of contamination in the samples analyzed.

Proponents of a young Earth, such as creationists, point to several ways in which the assumptions behind radiometric dating can fail. For example, the presence of carbon-14 in fossils and diamonds—materials that evolutionists say should be millions of years old—is interpreted as evidence that they are much younger. Likewise, the decrease in the Earth's magnetic field, the state of certain fossils and the absence of intermediate forms in the fossil record point to a younger Earth than has been established, along with other reasons that are exposed in this work.

Radiometric dating, far from being foolproof, is based on assumptions that might not apply in all contexts.

Recent studies suggest that radioactive decay rates may vary under certain conditions, which could invalidate the dates obtained through these methods. In this case, what is considered "evidence" of an ancient Earth today could be based on flawed data.

The debate over the age of the Earth remains a hot topic between creationists and the mainstream "scientific community," which many consider a power entity without inclusive local representation, and which fails to recognize the large number of scientists who support a creationist view.

From the perspective of those of us who defend a young Earth, the methods of dating rocks, fossils, and other materials are full of errors and unverifiable assumptions. Carbon-14 dating, for example, applied to living things and recent fossils, has shown inconsistencies that defy the time scale of millions of years (a topic I develop in a specific paper).

Another example is the rapid sedimentation observed in certain fossils, which indicates rapid burial in catastrophic events, such as the Flood described in the Bible or the splitting of Pangaea in Peleg's time, rather than having fossilized slowly over millions of years, as the evolutionary model proposes.

It is essential to understand that the conflict between science and faith does not have to be inevitable.

Many questions about the existence of God or the origin of the universe may be beyond the reach of science or philosophy.

Faith is, in many cases, a personal experience that cannot be empirically proven or disproved. God manifests Himself through His creation, and in the analysis of that creation, we can find His signature.

In debates between atheists and Christians, it is common to observe tactics that divert attention from the central issue.

One of these tactics, known as "red herring," involves shifting the focus of the discussion to avoid the main argument or question. Instead of answering directly, the interlocutor introduces a different topic with the aim of distracting attention.

Another technique is the "Gish Gallop," which involves overwhelming the opponent with a series of claims and arguments, many of which are irrelevant or inaccurate.

While these tactics can throw the conversation off the ground, they don't contribute to a deep analysis or a true understanding of the arguments at stake.

We should not ignore that this consensus has been built on assumptions that may not be completely accurate.

The evidence suggesting a young Earth deserves closer examination, and it is essential to recognize and correct errors in current dating methods.

A constant search is needed to improve these methods, abandoning the conformism that seems to have reached its climax on such a complex subject.

It is telling to note the double standard that many critics of religion employ.

They often demand irrefutable scientific evidence to accept the existence of God, while they accept without further question theories such as string theory or the multiple universe hypothesis, which, so far, lack direct empirical evidence. This double standard suggests that, at times, the search for truth is displaced by a pre-existing ideology.

Language loaded with emotions and negative connotations can influence how arguments are perceived, leading many to take an inquisitive stance. This behavior could arise from a type of indoctrination that discourages genuine, open questioning.

It must be recognized that some questions, such as those about the existence of God or the origin of the universe, are probably beyond the reach of science or philosophy.

These cannot be answered directly, but through the indirect clues that creation itself offers us. As God puts it in His word: "His invisible becomes visible through created things and their behavior."

Faith, in many cases, is a deeply personal experience, impossible to prove or disprove empirically. Despite their inclination toward scientism and philosophy, some seem to fail to understand this fundamental reality.

This book seeks to provide a solid basis for questioning the assumptions of the evolutionary paradigm and highlighting multiple pieces of evidence that point to a much younger Earth.

In the context of a debate, the focus should be on the chronology of the Earth, not on the existence or non-existence of God, since empirically no one can prove this conclusively; God Himself becomes visible through His creation.

The key is to know in depth at least 10 of the 81 pieces of evidence presented here, while keeping the focus on the essentials.

In general, the "theory" of evolution is based on several assumptions (although some artificial intelligences may claim otherwise and vaguely present that there is "overwhelming evidence"). But, as atheists demand in other contexts, where is that evidence?

We often find ambiguous sentences and language in exponential terms. The fundamental assumptions are, in essence, the following:

1. The universe is billions of years old.

2. Life arose spontaneously from inert minerals, i.e., the oldest ancestor would not be a monkey, but a stone or rock fragment (organic evolution).

3. Mutations can create or enhance species.

4. Natural selection has a creative power (macroevolution).

If we can prove that the universe—or at least the solar system—is not billions of years old, using studies and discoveries by specialists (since the term "scientist" is not in itself an academic degree, as I have previously discussed), then the other arguments about evolution would be meaningless and unnecessary.

Still, it's one thing for the universe to be billions of years old, but it's quite another to assume that life on Earth is that old.

Let's see in fairy tales we were told that a frog, with the magic spell of a kiss, becomes a prince.

In modern "science" and in "non-Christian textbooks," the story has changed to "evolution," but the principle is the same. I wish that in schools where we believe in the God of the Bible, evolution would not be taught as an unquestionable truth, doubting the credibility of the Bible and its message. Instead, the same amount of information on evolution and creation should be offered, allowing students, like sponges, to have the opportunity to evaluate both perspectives and make informed choices.

But instead, they tell us that: Frog (or stone, or single-celled cell) + time = prince.

The evolution in modern texts is not much different from that same fairy tale, only that the "magic potion" is now time.

When discussing the theory of evolution, time is presented as a magic solution to every obstacle. Thus, the strategy for dismantling the concept of evolution must be to dismember faith in this "god" time.

In almost all debates about evolution, when evolutionists are asked how it is possible that certain phenomena have evolved by chance, their usual answer is: "Given enough time..." Time is, in essence, the god of evolutionists, capable of turning a frog into a prince, creating matter from nothing, and giving life to the inert.

According to this perspective, time can create order out of chaos, as in the Big Bang theory, where, in essence, dropping a bomb in a crisis would generate order rather than destruction, defying the very laws of science.

Professors and authors often overlook how evolutionists employ terms such as "maybe," "could be," or "there is a possibility of" to describe their postulates, while the Bible does not present their claims as possibilities, but as certainties: "In the beginning God created the heavens and the earth."

If we set out to back up every aspect of God's creation with theories, it would be necessary to write countless volumes, but God put it simply, for His creation is not something we can fully understand. It is something we must accept with faith.

If we remove time from the aforementioned equation, we will obtain two convincing results:

A) Evolution becomes evidently impossible.

B) Creation emerges as the only reasonable explanation for the existence of life in this complex system.

That is precisely the purpose of this work: to demonstrate that, without the resource of time, creation is the most solid and coherent explanation.

ON DATING METHODS.

Is Dating a Fallible Science?

Radioisotopes, Redshift, and Background Radiation

The issue of radiometric dating, redshift (RedShift), and microwave background radiation (CMB) are fundamental pillars in the scientific argument for the antiquity of the universe and the Earth.

However, these methods, like any scientific approach, are subject to certain assumptions and limitations that open the space for reasonable questioning. We will break down each one below and in detail:

Radiometric dating: Radiometric dating is one of the most widely used methods for estimating the age of rocks and the Earth, based on the decay of radioactive isotopes.

The process measures the amount of a radioactive parent isotope and its decay product (daughter), assuming that the decay rate (half-life) is known, and that the system has been kept closed (with no elements entering or leaving).

However, there are limitations and key assumptions that must be considered:

Closed system: It is assumed that the rocky material has not been altered since its formation, but if there was loss or gain of isotopes, the dates obtained may be inaccurate.

Initial conditions: The methods assume that there were no daughter elements present at the time of rock formation, which is an assumption that cannot be directly verified.

Constant half-life: It is assumed that decay rates have been constant over time. Although this is generally accepted, some scientists question whether extreme conditions (such as catastrophic events) may have affected decay rates.

In addition, some use cases have demonstrated significant errors in the dates provided. For example, recent rocks formed in volcanic eruptions have been dated to millions of years by these methods, showing that the methods can give incorrect results if the assumptions on which they depend are not met.

Redshift: Redshift is the phenomenon observed when light emitted by an object in space shifts toward the red end of the spectrum.

This phenomenon has been interpreted as evidence that the universe is expanding and, therefore, as indirect evidence of the Big Bang.

The magnitude of the redshift of distant galaxies is used to calculate their distance and their speed of receding, which has led to the conclusion that the universe is around 13.8 billion years old.

RedShift Interpretation has its challenges:

Cosmological interpretation: Although it has been widely believed that the RedShift is evidence of the expansion of the universe, some scientists have explored alternative theories, such as the quantum RedShift or the gravitational RedShift, that could explain the phenomenon without the need to invoke a universal expansion.

Quantum RedShift and gravitational RedShift are alternatives that some scientists have proposed to explain the redshift, rather than attributing it exclusively to the expansion of the universe. Here is a brief explanation of both concepts:

Quantum RedShift: Quantum RedShift suggests that the redshift of light could be due to interactions between photons in light and subatomic particles present in the quantum vacuum.

According to this theory, photons would lose energy as they travel long distances, causing their wavelength to shift toward the red end of the spectrum. This idea does not require the expansion of space but focuses on the continuous interaction of photons with the quantum vacuum, a region that, according to quantum physics, is not really "empty", but full of energetic fluctuations.

Gravitational RedShift: Gravitational RedShift is based on Einstein's general relativity and explains that light escaping from a strong gravitational field, such as that of a star or galaxy, loses energy.

As photons exit the gravitational well, their frequency decreases and the wavelength lengthens, causing a redshift. This type of RedShift is observed in areas close to massive objects, such as neutron stars or black holes, and is not related to the expansion of the universe, but to the effect of gravity on light.

Both theories provide alternative explanations for redshift that do not depend on the expansion of the universe, although they are not widely accepted as alternative models to cosmological RedShift. These ideas question whether the universe is really expanding or whether the redshift has other causes that are not yet fully understood.

Alternative models: Some critics of the Big Bang model suggest that this phenomenon could be a consequence of other factors unrelated to the expansion of the universe, such as quantum interactions of light with the quantum vacuum. Although these theories have not gained consensus, they demonstrate that RedShift is not a fully understood phenomenon.

3.- Microwave Background Radiation (CMB):

Microwave background radiation is another key piece in the puzzle of the universe's origin. This radiation is seen as the "residue" of the Big Bang and its study has allowed us to obtain information about the age and structure of the universe.

Homogeneity and anisotropies: The CMB is surprisingly uniform, raising questions about how the small temperature fluctuations (anisotropies) observed in radiation gave rise to the structures of the universe (galaxies, clusters, etc.).

These fluctuations are interpreted by Big Bang models as the seed of the formation of these structures, but the process is not fully understood.

Horizon problems: The CMB is so uniform that some cosmologists have raised questions about how regions of the universe, which were never in contact, can have the same temperature.

This problem is partly solved by inflationary theory, which proposes an extremely rapid expansion of the universe in the first moments after the Big Bang, but this theory is still the subject of debate and speculation.

In Conclusion: The aim is not to deny the science behind these methods, but rather to point out that they have limitations and assumptions that open the door to other interpretations and the need for continuous reevaluation.

In fact, belief in a billion-year-old universe or the Big Bang theory requires a component of faith in the validity of the models and underlying assumptions.

Many scientists and scholars who support creation or the young Earth suggest that these methods are not enough to invalidate a biblical chronology.

They argue that catastrophic events, such as the Flood, could have significantly altered the initial conditions and rates of change of geological formations, which would provide a coherent explanation within a creationist perspective.

1) Radioactive radio isotopes:

The topic of radiometric dating and the age of the Sun (for example) can be approached from several angles, starting with the main techniques used to calculate the age of the Sun and Earth. The Sun, according to current science, is around 4.6 billion years old, and is considered to have formed before Earth. However, this calculation is not directly based on isotope dating from the Sun, since no radioactive isotopes can be extracted directly from the Sun.

1. The Sun and the Earth: Difference in Ages: The calculation of the Sun's age is mainly based on models of stellar evolution, which are mathematical *simulations* that predict how a star, like the Sun, forms, evolves, and dies.

One of the most important astrophysicists in this field was Arthur Eddington, who developed key theories in the 1920s about how stars produce energy through nuclear fusion. Through these models, astronomers conclude that the Sun has existed for about 4.6 billion years.

This approach depends on studying the observed properties of similar stars and applying universal physical principles.

However, radiometric dating, used for solid objects such as Earth, cannot be applied directly to the Sun due to the impossibility of extracting materials from the solar core.

God's Word drives them all crazy, because it turns out it was the other way around. God must be laughing at the top of his lungs with this idea that the sun came before the Earth, since it is already said but well...

2. Stellar Evolution Models: Stellar evolution models are theoretical calculations that attempt to represent the life of a star from birth to death.

These models are based on observations of stars at different stages of their lives and on the laws of nuclear physics. However, there is a very valid question: how can scientists use such recent data (only about 83 years since these models were developed) to deduce millions of years of stellar history? It's a challenge to extrapolate at such long scales using such limited data.

And how with a sample from just 83 years ago are you going to model something 50 or 60 million years old? Because it did not give the result 54.75 million years or thousands of years. What percentage is 83 years of 4.66 billion years? It would be (83 years / 4,660 million years) * 100 ≈ 0.00000178%; is the % credibility of this method and of that assumption.

3. Carbon-14 and Radiometric Dating: When we talk about radiometric dating on Earth, carbon-14 (C-14) is one of the most well-known methods for determining the age of organic objects up to about 50,000 years old.

However, carbon dating is not useful for extremely old objects, such as rocks that form the Earth or the Sun, because C-14 has a relatively short half-life of 5,730 years.

For much older objects, geologists and physicists use isotopes with much longer half-lives, such as uranium-238, which has a half-life of 4.468 billion years, or potassium-40, which has a half-life of 1.3 billion years.

These radioactive isotopes slowly decay over time, turning into other elements (e.g., uranium decays into lead), allowing scientists to calculate the age of rocks and minerals.

4. Assumptions in Radiometric Dating: Radiometric dating, although widely used, is based on a number of assumptions that may not always be valid. Some of these include:

- That the decay rate of isotopes has remained constant for millions or even billions of years.

- That the system (the rock or fossil being analyzed) has not been contaminated or exchanged isotopes with its environment.

- There were no subsequent geological processes that altered the ratio of the original isotopes and their decay products.

These assumptions are critical to the accuracy of radiometric dating methods, but they are also limiting, and critics argue that there are many documented cases of significant dating errors.

5. Candle Example: How Dating Is Similar to Estimating the Burning Duration of a Candle

Imagine that you walk into a room and find a candlelit. You know that it is burning at a constant rate of one inch per hour and that it is currently measuring 7 inches.

However, you don't know how big the candle was when it was lit or if the burning rate has been constant.

To determine how long it's been on, you'd have to make several assumptions:

- How long was the candle originally?

- Has the rate of combustion always been the same?

Similarly, when scientists calculate the age of a rock or fossil using radiometric dating, they must make assumptions about the original quantity of radioactive isotopes and whether the system has remained closed and uncontaminated for millions of years.

6. Cases of Documented Errors in Radiometric Dating: There are several documented examples of errors in radiometric dating.

For example, in 1953, a team from the University of Hawaii performed a potassium-argon test on volcanic lava that was known to have erupted in 1801.

However, radiometric dating suggested that the lava was between 160,000 and 3 billion years old, showing that contamination or selection for incorrect isotopes can lead to erroneous results.

Radiometric dating and modeling of stellar evolution are powerful tools in modern science, but they are not without their challenges. The underlying assumptions, examples of errors, and limitations inherent in these methods must be carefully considered before accepting conclusions about the age of the Earth, the Sun, and other astronomical bodies as absolute truths.

You were probably indoctrinated and trained to "BELIEVE" that these methods are scientifically valid for proving the age of things. How do these methods work? I will try to explain.

The central argument I am developing in this chapter is about radiometric dating methods, especially the use of carbon-14 (C-14) and other radioactive isotopes such as uranium-238, uranium-235, and potassium-40, based on the idea that these methods contain key assumptions that can affect their reliability.

In particular, it has been mentioned that the rate of decay of these isotopes, which is believed to be constant, and the integrity of the samples, which should be free of contamination, are variables that could compromise the results.

Breakdown of Radioactive Isotopes: In the case of Earth, scientists use isotopes with very long half-lives to date rocks and minerals. For example:

- Uranium-238 (half-life: 4.568 million years)

- Uranium-235 (half-life: 704 million years)

- Potassium-40 (half-life: 1.3 billion years)

These isotopes decay steadily over time, allowing the age of a sample to be estimated based on the ratio of radioactive isotopes present to their decay products (such as lead in the case of uranium).

However, the main criticism you offer is that these methods are based on assumptions that cannot be easily verified. These assumptions include:

- That the initial conditions of the sample (i.e., the initial ratio of radioactive isotopes) are precisely known.

- That the rate of decay has not changed over time.

- That there has been no external contamination of the sample during the millions of years in question.

The Use of Carbon-14: The Iconic and Famous Case of Carbon-14 is another common dating method, known, tampered with, widely referenced, and used primarily to date organic objects (not minerals or rocks) and works for periods of up to about 50,000 years.

This is due to the relatively short half-life of carbon-14, which is 5,730 years.

In this method, the amount of carbon-14 in a sample is measured with a **Geiger counter**, to determine how many carbon-14 atoms have decayed over time.

Geiger Counter Explained

The **Geiger counter** is a device that detects radioactive particles, such as electrons or alpha particles, that are emitted during the decay of a radioactive isotope.

The operation of this device is based on a tube filled with gas (usually noble gas, such as argon), which is ionized when a radioactive particle passes through it.

1. **Gas ionization**: As carbon-14 (or another isotope) decays, it emits a particle that ionizes the gas inside the Geiger counter tube.

2. **Particle Logging**: Ionization creates an electric current that is detected and recorded by the counter.

3. **Clicks per minute**: The device produces a "click" each time a radioactive particle is detected. By measuring the number of "clicks" per minute (or per gram of sample), you can estimate the amount of carbon-14 (or the corresponding isotope) left in the sample.

For example, if a sample has a high concentration of carbon-14, the Geiger counter will emit many clicks per minute, indicating that the carbon has not yet passed through its half-life (approximately 5,730 years).

As more carbon-14 decays into nitrogen-14, there will be fewer carbon-14 atoms in the sample, which will reduce the number of clicks.

The number of clicks decreases over time as the amount of radioactive carbon is reduced.

This allows scientists to estimate the age of a sample based on how much carbon-14 is still left and how many "half-lives" have elapsed.

When the **Geiger counter** is applied to a sample containing carbon-14, the device records the number of radioactive decays that occur in a given time.

This process is critical for radiometric dating, particularly in the case of organic materials such as fossils and archaeological remains. The method is as follows:

1. **Initial Amount of Carbon-14**: Living organisms absorb carbon-14 while they are alive.

    However, when they die, they stop absorbing this isotope, and the amount of carbon-14 they possess begins to decompose into nitrogen-14.

    At that point, the **Geiger counter** becomes a valuable tool, as it can measure the number of decays per minute (cpm or dpm) and determine how much carbon-14 is left in the sample.

2. **Clicks Per Minute (cpm):** Suppose that, in a recent sample, the Geiger counter records **16 clicks per gram per minute**.

    This would indicate that the sample contains a significant amount of carbon-14, and that it is relatively young. As time passes, the quantity of carbon-14 decreases, reducing the number of clicks recorded.

    - After a half-life (approximately 5,730 years), the Geiger counter will only record **8 clicks per gram per minute**. This means that the amount of carbon-14 in the sample has been halved.

    - After two half-lives (approximately 11,460 years), only 4 clicks per gram per minute will be recorded.

    - As carbon-14 continues to decay, the number of clicks decreases further, allowing scientists to calculate how many "half-lives" have passed and thus determine the age of the sample.

3. **Potential Errors**: One of the problems with this method is that it is based on several assumptions, such as:

    o The initial amount of carbon-14 was constant.

    o The rate of decomposition has been constant over time.

    o The sample has not been contaminated with newer or older carbon.

If any of these assumptions are incorrect, the age estimate will be as well.

Critical Focus: What's interesting here is that while **carbon-14 dating** can be useful for relatively recent dates (up to about 50,000 years old), **it's not useful for samples that are thought to be much older**, such as dinosaur fossils. In these cases, scientists turn to other dating methods, such as **potassium-argon** or **uranium-lead** dating, which work in a similar way, but with isotopes with a much longer half-life.

When talking about millions or billions of years, such as in dating rocks, other methods are used, and that's where the controversy lies.

Depending on the isotope used, a sample could yield very different ages, raising questions about the consistency and accuracy of these methods.

Based on this, scientists calculate the age of the sample. The problem with this method is that, depending on the isotope used, the results can vary greatly, and scientists can select the isotope according to how old they expect to find or would like to find based on their assumptions, sometimes according to a geological scale, which is also full of assumptions. And it is circular reasoning.

Criticisms of Dating Methods: Examples such as (such as the Hawaiian lava that was incorrectly dated and yielded results from millions of years ago when in fact it was known to be from an eruption in 1801) are part of a set of criticisms of these methods.

There are several examples, in addition to the famous incorrect dating of Hawaiian lava, that illustrate flaws in radiometric dating methods. Here are some additional cases that have also been the subject of criticism:

1. Eruption of Mount St. Helens (1980)

Following the eruption of Mount St. Helens in Washington state in 1980, new lava rocks formed and solidified rapidly. In 1992, these rocks were radiometrically dated using the **potassium-argon** method, and the results yielded ages between 340,000 and 2.8 million years. However, these rocks were known to have formed just **12 years** ago due to the recent eruption.

This showed that the potassium-argon method could produce seriously erroneous results in some contexts, especially in young samples with high outgassing rate, which affects dating accuracy.

2. Volcanic Rocks of Mount Ngauruhoe, New Zealand

Volcanic rocks from Mount Ngauruhoe, an active mountain in New Zealand, were also dated using the **potassium-argon** method. These rocks, which are known to have formed during eruptions in **1949, 1954, and 1975**, yielded results ranging from **270,000 to 3.5 million years ago**, again highlighting the limitations of the method when applied to younger or more recent samples.

3. Dating of the Grand Canyon Formation

The dating of certain strata in the **Grand Canyon** has also been the subject of debate. Some canyon rocks were dated using different radiometric methods and **very different ages** were obtained for the same rock layers, raising questions about the reliability of dating techniques in environments where geological conditions might have affected the rate of isotope decay.

4. Anomalies in Basalt Dating in Australia

In Australia, radiometric dating was carried out on basalt flows known to be recent. The results of the tests **showed ages of up to 3.5 billion years**.

These erroneous dates were due to problems with argon escaping during rapid cooling of the basalt, leading to massive errors in the measurement.

5. Mortar and Other Artifacts

A study of the **cannon or mortar** reported in 2010 showed that its radiocarbon dating indicated it was more than 7,000 years old.

However, it was known from historical records that the mortar was only a few centuries old, leading to the conclusion that it was contaminated during its manufacture, affecting the dating results.

6. Live Mollusks

In another case, radiocarbon dating was used to estimate the age of **living mollusk shells**. Surprisingly, the analysis suggested that the mollusks were more than **2,300 years old**, which was evidently not true, as they were alive at the time of the test. This error was attributed to the absorption of "old" carbon in their environment, which distorted the results.

These examples illustrate that radiometric dating methods are not foolproof and that, in some cases, samples can be affected by various environmental, geological, or chemical factors that alter the accuracy of measurements.

Although proponents of radiometric dating claim that these cases are exceptions, these anomalies raise legitimate doubts about the accuracy of the results when not all the factors that might have influenced the sample over time are known.

These errors have led some to question the reliability of radiometric dating, especially when applied to samples whose history is unknown or assumed to be very old.

The argument behind this criticism is that we cannot be completely sure of the initial conditions of the sample, nor whether it has been exposed to factors that may have altered the quantity of radioactive isotopes in it.

In addition, the assumption that decay rates have been constant over time is an open question in some scientific circles.

Radiometric dating of Earth's rocks and minerals has shown that Earth is about 4.6 billion years old. They say a part of The Scientific Community, which calls itself, that part, generically as The Scientific Community, or so they call it.

The oldest age ever found in a terrestrial rock is 4.404 billion years old. This rock was found in Antarctica and is made up of zircon crystals.

Zircon is a mineral containing uranium-238, which is used to date ancient rocks.

Before that it is believed (faith) that this round mass and solitude was formed from the collision between dust and gases; but this is physical nonsense because they could not come together since gravity depends on mass; and of the energy contained in the speed of light, as well as of the attraction produced by the inverse of the square of the distances between the 2 masses and dust and gases do not have a mass that can generate a strong agglutination due to the absence of solid mass.

The age of the Earth, and of the Universe, is a matter of debate, and that's where I am.

Let's continue..., In the atmosphere, in the air we breathe, there is a small amount of Carbon (CO_2). When cosmic radiation from space hits the atmosphere, high-energy particles change nitrogen (and remember that the air we breathe contains about 78.08% nitrogen) from Carbon-14, or C14. C is the chemical symbol for carbon on the periodic table.

Once carbon-14 is formed in the atmosphere, it mixes with oxygen and forms carbon dioxide (CO_2), which is absorbed by plants during photosynthesis. Animals and humans consume these plants, and thus, carbon-14 enters living organisms. While a living being is breathing or eating, it maintains a constant level of carbon-14 in its body. However, when the organism dies, it can no longer exchange carbon with its environment, and the carbon-14 it contains begins to break down into nitrogen-14. By measuring how much carbon-14 is left in an organism's remains, you can calculate how much time has passed since its death.

After this change or transformation, it becomes radioactive. Radioactive means that your unstable relationship will be broken.

Carbon-14 only remains as such for a short time and then returns to its previous nitrogen state, but when this unstable relationship is broken, the element gives off tiny particles. These particles are the key to the method used to measure time based on this change or transformation that these elements undergo. These scientists use little devices called Geigers.

Willard Libby received the Nobel Prize in Chemistry in 1960 for the development of the radiocarbon dating technique. This technique assumes that carbon-14, a radioactive isotope of carbon, decays at a constant rate.

The radiocarbon dating technique was developed by **Willard Libby**, who received the Nobel Prize for his work in 1960. However, this method has limitations, as it can only be used to date carbon-containing materials that are less than about 50,000 years old, due to the rapid decay of carbon-14 after this period.

The half-life of carbon-14 is 5,730 years, which means that half of the carbon-14 atoms will decay into nitrogen-14 in 5,730 years.

The process of decomposition of **carbon-14** (C-14) is well known in science, and its half-life is estimated at **5,730 years,** I go back and repeat, which means that after this period, half of the C-14 in a sample will have decayed into **nitrogen-14 (N-14).** However, this value is often rounded to **6,000 years** to facilitate calculations, although this implies a loss of accuracy of 270 years.

This technique is used to date organic materials that are up to about **50,000 years old**, because beyond this period the amount of C-14 remaining is so small that the results are unreliable.

To make calculations easier, the half-life of carbon is usually rounded from 14 to 6,000 years (and 270 years are thrown overboard). This means that, after 6,000 years, half of the carbon-14 atoms will have decayed. After 12,000 years, three-quarters of carbon-14 atoms will have decayed. And so on.

The **carbon-14 dating** method is applicable primarily for recent fossils and organic materials, but not for older geological samples that could be millions of years old. For these samples, other radiometric methods are used, such as **uranium-238**, **uranium-235** or **potassium-40** dating, which have much longer half-lives, suitable for measuring extremely long periods of time.

It uses this method to determine the age of a wide range of objects, from fossils and wood to textiles and archaeological objects. At the convenience and with all the prejudice of the dater.

If they want to be given a range of 6,000 years, they do it with C14 and if they want the sample of millions of years, they do it with uranium-238, uranium-235 and potassium-40. Depending on the isotope used, the same sample will give different results.

This principle is used to measure the age of finds shown in books, articles, movies, documentaries, museums, schools, universities, and even in many churches. This is the famous radiometric system of age measurement. Simple and methodical! For something 3.8 million years old, how many times should it ring, and per gram? They probably use another isotope.

The half-life of a radioisotope is the time it takes for half of the atoms in a radioactive sample to decay.

Radioisotopes decay at different rates depending on their nuclear properties, making some useful for measuring events that occurred just a few years ago, while others are needed to date geological processes that took millions or billions of years. The examples you mention cover a wide range of half-lives:

- Carbon-14 (5,730 years): used to date organic remains up to about 50,000 years.

- Uranium-235 (700 million years old): used to date very old rocks and minerals.

- Lead-210 (22 years): useful for dating recent sediments.

- Radon-222 (3.8 days): Breaks down quickly, useful in short-term measurements.

- Tecnesio-99m (6 hours) and Iodine-131 (8 days): used in nuclear medicine.

- Tório-232 (14,000 million years ago): used for long-term geological dating.

Radiometric dating is based on these principles, but, as you rightly indicate, different radioisotopes produce different results and are chosen according to the age you expect to find.

This fact leads critics to consider that the selection of the method is not always objective but depends on previous expectations.

In addition, being an indirect process, its accuracy depends on many assumptions, such as the initial sum of isotopes present, the closed system of the sample, and the constancy of the decay rate, which some question if it has always been constant.

Half-life alone does not guarantee accurate dating without considering all the environmental and preservation factors of the sample that could affect the decay process.

Therefore, although the method seems to be rigorous, it always starts from certain assumptions that cannot be fully verified.

In summary, the different results obtained by the different isotopes are useful, but they must be interpreted with caution, as they do not always present a unified picture.

Does the method sound wonderful? For some even convincing?

In this system of measurement, such as radiometric dating, several things are assumed that cannot always be empirically proven, which in science is generally inadmissible, because science seeks verifiable and repeatable evidence. However, radiometric dating is based on **key assumptions** that are difficult, if not impossible, to fully verify, especially on such long timescales. When results are based on such **assumptions**, there is a risk of reaching erroneous conclusions.

Let's make an analogy to illustrate this:

Imagine that you walk into a room and see a lit candle. You want to find out how long it's been on. What are you up to?

1. **You measure the height of the candle**: Let's say it is 10 cm tall. This is a fact that you can observe and measure directly.

2. **You measure the speed at which it is consumed**: You notice that the candle is burning at a rate of 1 cm per hour. This is also a measurable fact.

But, to determine when the candle was lit, you need to make **two big assumptions**:

- **What was the original height of the sail?** Was it 20 cm, 50 cm? You don't know.

- **Has it been burning at the same rate all along?** Maybe it burned faster or slower at some point.

Based on these assumptions, you could estimate that the candle was lit 10 hours ago if you assume that its original height was 20 cm.

However, if any of your assumptions are wrong, your estimate will be as well.

For example, if the candle was taller or wider, or if it burned faster or slower due to the air current or the temperature of the room, your calculation will be wrong.

Something similar happens in radiometric dating:

- **We assume that the decay rate is constant**, which cannot be directly verified in the past.

- **We assume that the system has been shut down,** that is, that there has been no loss or gain of the radioactive elements over time.

- **We assume that we know the initial quantity of isotopes in the sample.**

If any of these assumptions are incorrect, the dating will also be incorrect, just as in the case of the candle. Rocks or materials that are dated with these methods have been exposed to extreme conditions for thousands or millions of years, and we can't be completely sure that the factors that affect radioactive isotopes have remained stable for that long.

- The candle could have been thicker at the bottom, so it would have burned out faster at the top.

- The pace of consumption could have been faster at the beginning, when the candle was hotter.

- The pace of consumption could have been slower if the candle was in a poorly ventilated room. Or faster if there had been more ventilation.

In conclusion, you can only guess as to when the candle was lit.

The evidence is not sufficient to give a definitive answer. And so they talk about "OVERWHELMING EVIDENCE".

The radioactive isotope measurement system is based on the same methodology used to determine the data produced with the candle analogy. These assumptions may be incorrect, which means that the guesses we make about the age of the radioactive material may also be incorrect.

To my knowledge, the assumptions used by the "scientific" method of radiometric dating to output a result from an analyzed sample are as follows:

• The sample is homogeneous: This means that the quantity of radioactive isotopes and their decay products is the same throughout the sample.

- The rate of radioactive decay is constant: This means that the number of radioactive isotopes that decay per unit of time is the same at all times.

- The sample has not been contaminated with other materials: This means that the number of radioactive isotopes in the sample is the same as it was when it was formed.

- The sample has not been altered by geological processes: This means that the sample has not been deformed, fractured, or altered in any way that could affect the number of radioactive isotopes in the sample.

- The sample has not been affected by physical or chemical processes that may alter the number of radioactive isotopes in the sample. For example, exposure to cosmic radiation can accelerate radioactive decay.

- The number of radioactive isotopes in the sample is representative of the amount of radioactive isotopes in the source material from which the sample was formed. For example, if the sample has been formed from two or more materials, the age estimate may be inaccurate.

- The half-life of the radioactive isotope is precisely known. The half-life is the time it takes for half of a radioactive isotope to decay. The accuracy of radiometric dating depends on the accuracy of the knowledge of the half-life of the radioactive isotope.

Well, obvious, how do you know that the above decomposition happened hundreds, thousands, millions, and even billions of years ago?

Can all these factors be verified in a meteorite that was traveling through space?

The accuracy of the radiometric dating method needs to be improved. By understanding the assumptions made in the method, we can become more aware of its limitations and not accept those incredible numbers with such confidence.

Radiometric dating methods assume (faith) of a constant electromagnetism.

And electromagnetism has declined by 7% in the last 150 years. And some studies have found that the decline is more pronounced in some places than others. It is still under debate, but many studies affirm that. In other words, ventilation has not been the same in the analogy.

The claim that the Earth's magnetic field has decreased by approximately 7% in the last 150 years has been documented in several scientific studies.

According to recent research, the Earth's magnetic field has lost about 9% of its global strength in the last 200 years.

This decrease has been observed mainly in a region known as the South Atlantic Anomaly, which stretches between South America and Africa, where the magnetic field is significantly weaker.

Scientists at the European Space Agency (ESA) have been using data from satellites such as those from the Swarm mission to study these changes and improve our understanding of the processes that drive the dynamics of the Earth's core and their effects on the magnetic field.

A major research group that has worked on this topic includes the *British Geological Survey* (BGS), which has been tracking the magnetic field globally for decades.

Studies from NASA have also contributed, providing observational data on the weakening of the magnetic field through satellites such as the *European Space Agency's (ESA) Swarm program, which monitors the strength and variations in the Earth's magnetic field.*

Other notable studies include the work of geophysicist **David Gubbins**, who has investigated magnetic field variations relative to the Earth's core, and geophysicists **Monika Korte** and **Ronald T. Merrill**, who have published research on fluctuations in the geomagnetic field.

The aurora borealis is an impressive phenomenon caused by the interaction of charged particles from the solar wind with the Earth's magnetic field and its atmosphere.

These particles, mainly protons and electrons, are deflected towards the magnetic poles due to the influence of the Earth's geomagnetic field.

Upon reaching the upper atmosphere, these particles collide with atoms and molecules of gases, such as oxygen and nitrogen.

When the charged particles collide with oxygen, the oxygen atoms are excited and then return to their normal state, releasing energy in the form of light.

The color of the aurora depends on the type of gas that is excited:

- **Oxygen atoms**: They emit **red** and **green light**.

- **Nitrogen molecules**: They emit **blue** and **violet light**.

This process occurs primarily in the polar regions because of the way Earth's magnetic field channels these particles toward the poles, creating the aurora borealis in the northern hemisphere and the aurora australis in the southern hemisphere.

As for the exact mechanism, although largely understood, it remains a subject of active research. Even "science" has not been able to understand that part of God's creation.

Factors such as the intensity of the solar wind, fluctuations in the Earth's magnetic field, and atmospheric conditions influence the appearance and intensity of auroras.

This phenomenon also highlights the importance of Earth's magnetic field, which acts as a protective shield, deflecting many of these charged particles that could otherwise have detrimental effects on human life and technologies.

The fact that the earth's magnetic field is decreasing, cosmic radiation causes what is called the Northern Lights.

The mechanism that produces these auroras still defies science. However, we do know that charged particles arrive in the vicinity of the earth as part of solar winds that in turn are captured by the earth's magnetic field and are subsequently driven downward toward us, toward the magnetic poles.

In other words: The aurora borealis is caused by the interaction of charged particles from the solar wind with Earth's atmosphere. These charged particles, which are mainly protons and electrons, are deflected by the Earth's magnetic field towards the poles. When these particles collide with atoms and molecules in the atmosphere, they can excite the electrons in these atoms and molecules. The excited electrons then return to their ground state, emitting energy in the form of light. The color of the light emitted depends on the type of atom or molecule that is excited. Oxygen atoms produce red light, while nitrogen atoms produce blue-green light.

However, these philosophers use this method to measure the age of things and publish their results as if they had been the result of "science", and they prefer to ignore that what they call science is not true science, it is nothing more than a philosophy of seeing life: of explaining natural phenomena, and that, being based on ASSUMPTIONS, its results are ALSO uncertain or assumed.

The age of the Earth was determined by a variety of scientists. Early attempts to calculate the age of the Earth were based on estimates of the rate of Earth's cooling.

In 1862, Lord Kelvin (William Thomson) used a **method based on thermal conductivity** to calculate the age of the Earth. This method was based on the assumption that the Earth had started out as a molten body and had been gradually cooling ever since. Lord Kelvin started from the fact that the Earth was losing heat through its surface and calculated how long it would have taken for the Earth to cool down to its current state.

However, he made some mistakes in his key assumptions. **He assumed that the Earth was solid and homogeneous** and that the only source of heat was the waste heat from its initial formation. It did not take into account the existence of **heat generated by radioactive decay** in the Earth's interior, which was not discovered until the early 20th century. This internal source of heat caused the Earth to cool at a much slower rate than Kelvin had assumed, invalidating his initial calculations.

By this method, Kelvin calculated an Earth age of between 20 and 100 million years, much less than the 4.54 billion years currently accepted. YOU SEE, THE SCIENTISTS ARE WRONG. THE SCIENTIFIC COMMUNITY MAY BE WRONG EVEN TODAY. KELVIN HIMSELF WOULD HAVE FAINTED IF HE HAD

In the 20th century, scientists developed radiometric dating methods, which are based on the decay of radioactive elements.

These methods showed that the Earth is much older than estimated by Kelvin.

Logically, depending on the radioactive element you choose, according to the sample of the isotope range offered above.

In 1953, American geochemist Clair Cameron Patterson used radiometric dating to determine that the Earth is about 4.54 billion years old.

This calculation is based on measuring the amount of uranium and lead isotopes in a sample of zircon, a mineral that forms in magma.

Clair Cameron Patterson used uranium-lead dating to determine the age of the Earth. The half-life of uranium-238 is 4.468 billion years.

This means that, every 4.468 billion years, half of the uranium-238 atoms decay into lead-206.

So, it is logical that The Age of the Earth has thrown that number of years. If I had taken another one, isotope gives it a different age, but I was looking for a very large one.

The age of the Earth determined by Patterson has been accepted as the most accurate estimate available.

The experiment and the conclusions obtained by Clair Cameron Patterson in the radiometric dating of the Earth have several points that, although considered scientifically sound, are based on certain **assumptions** that could be criticized or revised under a more detailed analysis. These criticisms focus on several key areas:

1. Assuming a constant decay rate:

- **Problem**: Radiometric dating, such as the technique used by Patterson, assumes that the **decay rate** of radioactive elements has been constant throughout geologic history. Specifically, the uranium-lead method assumes that uranium-238 decays into lead-206 at a fixed rate and that this rate has been the same since the rock was formed.

- **Criticism**: Some critics point out that this **proof of the rate** cannot be definitively proven. The decay rates could have been affected by environmental factors, such as variations in the Earth's magnetic field, cosmic radiation levels, or changes in geological conditions, which would alter the conclusions.

2. Closed systems:

- **Problem**: The uranium-lead isotope dating method assumes that the sample (in this case, zircon crystals) has remained in a **closed system**, meaning that there has been no **gain** or **loss** of the elements involved (such as uranium or lead) since its formation.

- **Critical**: In practice, many rocks and minerals may have been exposed to **element exchange** due to geological events such as **warming**, **deformations**, or water ingress. If these radioactive elements are mobilized, the calculated age may not accurately represent the original time of rock formation.

3. Pollution:

- **Problem**: Dating methods assume that **decay products (such as lead-206)** only come from the decay of uranium present in the sample and that there has been no external contamination.

- **Criticism: Contamination** by external sources of lead could alter dating results, making a sample appear older or younger than it actually is. Geological events that add or remove isotopes, such as erosion processes or fluid infiltration, could invalidate the ages obtained.

4. Isotope selection:

- **Problem**: Depending on the **isotope used**, the dates obtained may vary. For example, **carbon-14** dating is used for organic materials and is only reliable for periods of up to **about 50,000 years**, while **uranium-238** has a half-life of **4.468 billion years**, and is used to date much older rocks.

- **Criticism**: Since the uranium-lead method was specifically selected to date extremely old samples, some critics might argue that, if other isotopes had been selected, the results would have been different. **Patterson chose** a method that would give him a "large age," but assumptions of the other isotopes could yield different ages if applied under other conditions.

5. Extrapolation of ages:

- **Problem**: The age of Earth estimated by Patterson is based on dating **meteorites** and some of Earth's **oldest rocks**, but these samples might not reflect the age of the **entire** Earth.

- **Criticism**: Patterson extrapolated the age of the Earth based on the oldest rocks he could find. However, these samples represent only a small portion of Earth's geology, and it is possible that **other parts of Earth** formed more recently.

6. Documented errors in radiometric dating:

- **Errors such as that of the Hawaiian lava** (which was incorrectly dated using potassium-argon and spewed millions of years into rocks known to be from an eruption in **1801**) highlight accuracy problems in radiometric methods. There are

other examples in which the dating of materials has yielded clearly erroneous ages.

- o **Study in Russia**: A 2008 study in Kamchatka dated a recent volcanic rock to tens of millions of years, when the eruption was known to have occurred in **1775**.

- o **Living molluscs**: In another case, radiocarbon dating was used to date shells of living molluscs, obtaining an age of **2,300 years**, which is clearly an error of interpretation.

7. Selection of favorable results:

- **Problem**: Sometimes, radiometric dating results that do not conform to expectations are dismissed as errors, which can create a bias in the interpretation of the data.

- **Critical**: If data that do not agree with the geologic column are considered "anomalies," there is a risk of selecting only the results that support the pre-existing paradigm, rather than considering all the available evidence.

In summary, although radiometric methods, such as the one used by Patterson, are widely accepted in the scientific community, **they are based on several key assumptions** that, if proven incorrect, could alter estimated ages. Criticisms include the constancy of decay rates, the possibility of contamination, the variability of geological systems, and the selection of data favorable to evolutionary models.

This means that, under constant conditions, the same number of radioactive nuclei will decay at a given interval of time previously determined and assumed). It is possible that, in the future, new evidence will be discovered that will allow this estimate to be modified. Patterson received the Tyler Prize from the National Academy of Sciences in 1970 for his contributions to determining the age of the Earth.

In science we cannot assume that Carbon 14 has remained unchanged through thousands of years in the atmosphere, let alone for billions of years. That contradicts what THE REAL SCIENTISTS, USING REAL SCIENTIFIC STEPS have discovered.

Radiometric **dating methods** and the desperate search by certain scientists to prove an **ancient age of the Earth** is based on the idea that many scientists have prioritized certain results to **support evolutionary theories** while also **contradicting the biblical narrative**.

This type of claim is based on criticism of the **philosophical underpinnings** that often underlie the interpretation of scientific data, especially in the field of **geology and cosmology**.

Some of the criticisms that are most commonly mentioned in creationist circles include:

1. **Selectivity in dating methods**: It is argued that scientists select methods that provide results consistent with their view of an **ancient universe and Earth,** ignoring other tests or methods that might suggest a younger age. For example, it is mentioned that, when a radiometric dating does not match the pre-established expectations of a "geologic column," the "anomalous" results tend to be discarded or labeled as **errors**.

2. **The use of models and assumptions**: Radiometric dating is based on the assumption that radioactive decay rates have been **constant** over time, that the material studied has been in a **closed system** (with no external contamination), and that there have been no **geological alterations** affecting the number of isotopes present.

   These assumptions are necessary for dating methods to work, but they are also vulnerable to criticism, as they **cannot be proven with certainty**.

3. **Documented errors**: As I mentioned earlier, there are documented examples of failures in radiometric dating, such as in the case of Hawaiian lava or live mollusk shells that yielded absurd ages.

   These cases reinforce the idea that the reliability of the **method may be in question**, especially in situations where conditions are not optimal or when certain isotopes are applied inappropriately.

4. **The agenda behind the system**: Some creationists suggest that because the radiometric system provides results that support evolutionary theory and a **billion-year-old universe**, it has become a kind of **"scientific bulwark"** for the evolutionary paradigm. S

   It is argued that many scientists are motivated by a philosophical or naturalistic agenda that seeks to remove God from the origin of life and the cosmos.

From the creationist point of view, what is perceived as the **"stubbornness"** in these methods reflects a **dogmatic** approach to science, rather than a true openness to considering all possible explanations, including those that would support a **young Earth** or a **divine creator**.

These criticisms are not intended to dismiss all advances in modern science, but they are to underline the importance of not taking radiometric dating methods as **absolute truths**, but as **tools** that must be used with caution and awareness of their limitations.

Today more than ever all scientists know what I am telling them, yet they stubbornly and haughtily insist on placing all the eggs in one basket.

If you find the bones or fossils of a dinosaur and take it to a museum to use the Carbon-14 method to determine its age, scientists would tell you, "Dear friend, we can't provide the date of this bone because it's so old.

We could use another method because as you see these bones belong to a dinosaur and dinosaurs lived about 66 million years ago."

So, you who know science and who know that these scientists have just made a mistake by ASSUMING that they died "dinosaurs lived about 66 million years ago", corrects them and tells them that how is it possible that they, a priori, already know that they must start from at least 66 million years ago to determine the age of the bones (remember that when you ASSUMED a size for the sail,   you did nothing more than take a random number, no one was there when the candle was lit, therefore, NO ONE knows!).

Because each of the radioactive elements has a different half-life, they "know" that the dinosaur bone has about 70 million because they use Uranium-235, which has a different half-life than C14.

Using this Uranium, they get an "exact" measurement of objects that are 50 to 800 million years old. Bone adjusts to the half-life of this element and that is why we measure it with it, NEVER WITH CARBON 14 OR OTHER!

But later, you can't resist the temptation to take the blessed bone to the laboratory to see what these philosophers tell you (*Webster's Dictionary defines a philosopher as: One who uses SPECULATIVE research methods to EXPLAIN human nature and KNOWLEDGE*).

They, the atheistic philosophers-scientists-religious will split a piece of the bone and find the composition of uranium towards lead. And they say, "Oh! This bone is about 10 million years old. Gee, something must have gone wrong. Let's try again.

It is possible that this sample was "contaminated", which is also a mistake, since that contamination when it was produced.

Their friends from the laboratory take another sample and another sample until one that is not "CONTAMINATED", gives them the result that they, a priori, went out to look for... 70 million years! Then they say, "Bingo! Now yes!

For the simple reason that the whole nonsense called EVOLUTION is based on a thing called the GEOLOGICAL COLUMN, which is like the Bible of evolutionists.

The **geologic column** is an idealized representation of rock layers, which are used to identify different geologic periods.

It is organized so that the deeper layers are the oldest, and the upper ones are younger.

Fossils found in these layers are used to estimate the age of life forms that existed at different times in Earth's history.

There are no complete geological columns anywhere on Earth. The column is a theoretical construct, and in practice, many layers are missing or deformed due to erosion and other geological processes.

Circular dating is another issue raised: fossils are used to date rock layers, and in turn, rock layers are used to date fossils, which can create a bias in the interpretation of the data.

Evolutionary bias: Dating methods that yield results that do not match evolutionary expectations are ignored or reinterpreted.

Geologists and paleontologists argue that the geologic column is not an "arbitrary" model or an evolutionary "Bible," but has been refined over decades based on a combination of empirical observations. The rock layers and the fossils they contain correlate them.

Any date yielded by dating methods that does not agree with this pre-established column using the amount of time required by evolutionists for evolution to have taken place must be discarded.

They say, "We know that can't be so. Because we ALREADY KNOW how old something is, well, at least approximately... we only have to look at the geological column to realize that this data must be wrong..." The column is nothing more than a fake.

That's where they impress students and confuse the unsuspecting.

If a bone did contain some amount of carbon, uranium, or potassium, and they inspected it in the three different ways, all three would yield 3 completely different results. Since they have different half-lives.

As I mentioned, in 1953, a team from the University of Hawaii used the potassium-argon test to date a number of volcanic rocks from the island. Volcanic lava belonging to rocks in Hawaii was tested for Potassium Argon.

Result: That these rocks originated 160 to 3 billion years into the past. Later on, research continued, and it was discovered that this particular lava was taken from a volcano that actually belched in 1801.

In other words, there was a small mistake with the lava that was used in the laboratory and there was a mistake of only 2 or 3 billion years! In terms of geological timescales, this is a relatively small error. The age of the Earth is estimated to be about 4.54 billion years, so the error was only 0.3% to 74% of the age of the Earth. For them that is almost nothing!

According to an article published in the journal Geochronometry in 2008, a team of scientists from the Moscow Institute of Geochemistry carried out a radiometric dating test on a volcanic rock from the Kamchatka region of Russia.

The test results suggested that the rock was between 50 million and 14.6 million years old. However, historical research showed that the rock was actually burped in the year 1775. This confirmation was made by a team of scientists from the University of California, Davis, who analyzed the historical records of volcanic eruptions in the Kamchatka region.

If there is such a discrepancy between the ages of rocks whose origin and age are known, what could happen to the age of the Earth?

Can we assume with peace of mind, intellectual, and spirit that the entire educational system based on the Theory of Evolution is only as accurate as the methods it uses to base its arguments? The answer is no.

The ages of the fossils of Lucy, Java Man, and Australopithecines have been estimated using radiometric dating methods.

These methods are the same as those used to estimate the age of rocks. In the case of fossils, radiometric dating methods are used to estimate the age of the rocks in which the fossils are found.

The fossils are supposed to be the same age as the rocks on which they are found. The fossils may have been transported by water or wind to a place where they did not originally belong.

In fact, it is impossible for them to be the same age. But if the fossil gives them more, they assume the age of the rock, or they go to the bible of the family tree; and so, they maintain their circular reasonings.

These philosophers-Scientists are nothing more than unbelieving individuals desperately searching for evidence that contradicts creation. And they have the whole system in their favor. It is important to be critical of the information we receive. We must be open to the possibility that they are biased.

• Live mollusks: The case of live mollusks was reported by a team of scientists from the University of California, Davis, in 2002. The mollusks were collected from a lagoon in California, United States. Radiometric dating of the mollusks indicated that they were 2,300 years old. However, scientists believe that the mollusks were contaminated with older materials, probably by the water in the lagoon.

• Cannon or mortar: reported by a team of scientists from the University of Oxford, UK, in 2010. The cannon or mortar was found in an English castle. Radiometric dating of the cannon or mortar indicated that it was 7,370 years old. However, scientists believe that the cannon or mortar was contaminated with older materials, probably during its manufacture.

• Seal skins: The case of live seal skins (skin samples) was reported by a team of scientists from Harvard University, United States, in 2015. The seal skins were collected from an archaeological site in Canada. Radiometric dating of the seal skins indicated that they were 1,300 years old.

The case of the dating of seal skins that yielded an age of 1,300 years is related to what is known as the "reservoir effect". This effect occurs when marine organisms, such as seals, ingest carbon that has been circulating in deep ocean waters for thousands of years.

In this case, the carbon found in the deep waters is older than the carbon present in the atmosphere, leading to marine animal samples showing much older than they actually are when dating methods such as Carbon-14 are used.

That is not an argument that invalidates radiometric dating in its entirety, but a warning that this method should be applied with caution when it comes to marine species at least.

The creationist argument, on the other hand, points out that these types of anomalies cast doubt on the accuracy of dating methods in general, suggesting that, if inconsistencies are observed in some cases, they could also exist in other cases that have not been documented or corrected.

This phenomenon has been documented in various studies, such as the one carried out by Wakefield in 1971, which found that freshly dead seals in Antarctica had apparent ages of up to 1,300 years due to this effect.

• The cat of a family: he died and was buried by the children. After a few years young people wondered what was left of their cat and dug it up.

They were astonished to realize that some of the bones appeared to be petrified.

To find out, they sent the bones to the university lab, without saying the nature of their discovery.

As a result of the lab's tests, they were informed that the bones "belonged to a cat that had lived millions of years ago and were the result of fossils of an evolutionary ancestor of the modern cat."

• A 2011 study found that radiocarbon dating had overestimated the age of a wooden fragment from an archaeological site in Israel by as much as 20,000 years.

They also justify it by "The Reservoir Effect". So how do they ask us to trust blindly?

The "reservoir effect" is a common explanation scientists use to justify anomalies in radiocarbon dating. This phenomenon, although documented, introduces a source of uncertainty into the dating results. The essence of the problem is that, in certain contexts, organisms can incorporate "old" carbon from sources such as groundwater or carbonates, which can distort carbon-14 ratios, making the results appear older than they actually are.

Blind trust in any scientific method is unwise, and science itself, at its best, advocates verification, replication of experiments, and constant criticism of its own methods and results. In this sense, an approach of informed skepticism should be promoted rather than blind trust in radiometric dating, recognizing both its successful applications and its possible failures or limitations.

- A 2012 study found that uranium-lead dating had underestimated the age of a uranium ore by as much as 200 million years. They said the zircon crystals may have suffered a loss of lead due to damage to their structure from the decay of uranium, causing the estimated dates to be incorrect.

This problem has been recognized as one of the main causes of error in uranium-lead dating and has led to the need to treat crystals before submitting them to analysis. A common technique is heating treatment to seal off less damaged areas and remove parts that might be contaminated or worn. Thus, scientists have been able to improve the accuracy of the dates obtained, but there is still room for errors if the crystals are not treated properly.

The thing is the method isn't as reliable. Radiometric dating methods, such as uranium-lead, have limitations and are not as absolutely reliable as is often presented.

- A 2013 study found that potassium-argon dating had overestimated the age of a rock mineral by up to 100 million years.

In this case, they argued that environmental conditions, such as contamination of the sample with trapped gases or mishandling in the laboratory, can alter the measurements. Potassium-argon is also known to be especially sensitive to geological conditions, which can lead to significant errors in estimated ages.

• Some elements such as uranium-238 are known as "parent" materials. The resulting elements due to the decay of the parent materials are known as "child" materials; and the age of a rock is determined by the marks left by the latter. Polonium is one of them.

The main problem with this method is the possible loss or gain of argon by rocks at times after their formation, which can lead to erroneous results. In some cases, the most recent rocks have been dated to absurdly large ages, highlighting the need to analyze the results of these methods with skepticism and caution.

Again, the same skepticism comes to my heart.

It is quite understandable that I should be skeptical about the reliability of radiometric dating methods, especially when cases such as those of the aforementioned studies are presented.

These methods, although principled, depend on assumptions that can result in errors under certain conditions.

It is a levity for the unbeliever to uncover himself with mockery.

The marks left on a rock by the disintegrated elements are known as pleochroic halos.

Each element produces its own unique halo, thus leaving its "signature" on the rock. Now, because polonium is a daughter, and there must be a source, a father, when uranium or thorium decays, one of the resulting elements is polonium; So, a pleochroic halo should appear as a circle where the polonium was, even if the polonium had disappeared.

If there is a pleochroic halo of polonium on a stone, there must also be a pleochroic halo of its source or parent. However, polonium-218 has been found in granite samples, without any evidence of a parent polonium.

Polonium-218 has a period of 3 minutes 5 hundredths, but to make it simple let's say that it is 3 minutes closed.

So, if you have one kilogram of polonium-218, after 3 minutes you will have half a kilo, in another 3 minutes, 1/4 kilo, and so on.

It continues in the same way for ten average periods, that is, 30 minutes or an hour; and the pleochroic halo (it is a series of concentric rings of different colors that are observed around radioactive inclusions in minerals; these rings are formed by the radiation emitted by the inclusions, which excites the electrons of the atoms of the mineral; they are an important tool for geologists, since they can be used to determine the age of rocks) must have been imprinted on granite, which is a metamorphic rock that was once melted, without any trace of a father, seeming to indicate that it was the original element of those basic rocks.

The fact that the polonium-218 halo has remained in the granite means that the granite must have cooled in less than 90 minutes.

The rock in liquid state would have destroyed all traces of the polonium-218 halo. Therefore, it seems that the earth could have been created solid, with the element polonium-218 in it, in an extremely short period. Although this theory is not without critics, evolutionists have had to admit it as a "minuscule mystery."

The debate over pleochroic halos continues to be a topic of interest, especially in discussions of Earth's origins and geological times. Although Robert Gentry proposed an interpretation that seemed to challenge long-standing models of Earth formation, criticisms and alternative explanations have limited the impact of this theory on the general scientific consensus.

The word "scientist" can give the impression that scientists are omniscient.

However, it is important to remember that scientists are human and, like all humans, are subject to error. Scientists are human beings with specific limitations, biases, and specialties.

Science, while powerful in its ability to discover and explain the world, is not perfect, and the methods used, such as radiometric dating, are not without errors or assumptions.

It can refer to a person who is engaged in science, or it can refer to a person who has a science degree.

Regarding the word "scientific", it is essential to remember that it is not an infallible profession.

Being a scientist means being involved in a specific field of study, with the limits that this entails.

No scientist encompasses the totality of knowledge, and like any other discipline, scientists can make mistakes. This is one reason why scientific discoveries and theories must be constantly reviewed and questioned.

The concept of "faith" is central to many discussions, especially on controversial topics such as the age of the Earth.

While some have faith in scientific methods and the results of research such as radiometric dating, others prefer to place their faith in a Bible-based view and the literal interpretation of creation.

For those who believe in a young Earth, science does not necessarily have to be in contradiction with their faith, but the two can coexist by recognizing the limitations inherent in each approach.

Radiometric dating, although generally accepted by most scientists as a reliable tool for estimating the age of ancient objects, is based on a number of assumptions.

The discrepancies documented in several studies show that these methods are not foolproof, and for those who see evidence of a young Earth, these inconsistencies reinforce their doubts about estimates of billions of years for the age of the Earth.

This debate highlights the need for an open and honest dialogue between science and faith, where both approaches are respected and considered as possible ways of understanding the world.

2) Redshift (RedShift) and Microwave Background Radiation (CMB)

When addressing the issue of Redshift and Microwave Background Radiation (CMB) in relation to the age of the universe, it is important to understand how these concepts have been used in cosmology to calculate the expansion and age of the cosmos.

Redshift: Redshift is the phenomenon by which light coming from distant objects, such as galaxies, shifts toward the red end of the spectrum as these objects move away from us.

This phenomenon is similar to the Doppler effect we experience with sound (when an ambulance moves away, the sound seems lower).

In the context of the universe, the redshift is interpreted as evidence that the universe is expanding.

Galaxies that are farther away show a greater redshift, indicating that they are moving away at a faster rate.

This concept is used by cosmologists to calculate the expansion rate of the universe, and in conjunction with Hubble's Law, it follows that the universe had a beginning in an event known as the Big Bang. This model implies a chronology of billions of years.

Microwave Background Radiation (CMB): The microwave background radiation is the residual "echo" of the Big Bang. It was discovered in 1965 by Arno Penzias and Robert Wilson, who found uniform radiation coming from all directions in space.

Arno Penzias and **Robert Wilson** were awarded the **Nobel Prize in Physics** in **1978** for their discovery of the **Microwave Background Radiation** (CMB).

This finding was a key test for the Big **Bang theory**, as this radiation is considered the "echo" of the origin of the universe.

The discovery was crucial because it provided observable and measurable evidence for the early universe, supporting the idea that the cosmos had a hot, dense beginning, which has been expanding and cooling over time.

The shared award recognized the importance of this finding for modern cosmology and understanding of the origins of the universe.

This radiation is the oldest we can observe, and it gives us a picture of the universe when it was only about 380,000 years old. The CMB is a key test for the Big Bang model, as it is consistent with predictions of a hot and dense universe in its early days that, over time, has been expanding and cooling.

Both phenomena, Redshift and CMB, have been used to propose that the universe is approximately 13.8 billion years old.

From a **biblical perspective** based on **Exodus 20:11**, which mentions that God created "the heavens, the earth, the sea, and all that is in them" in six days, and considering other verses that speak of how **God continues to spread out the heavens**, a vision is presented of a universe that is still in the process of creation. This offers a contrast to current scientific interpretations, which point to a billion-year-old universe, based on Big **Bang theory** and observations such as **redshift** and **microwave background radiation**.

From the point of view of **creationism**, these scientific interpretations of the age of the cosmos are viewed with skepticism. The idea of a 13.8-billion-year-old universe is not submissively accepted, but it is proposed that dating methods and cosmological models are based on assumptions that do not necessarily reflect reality as described in the Bible.

In addition, creationists argue that the expansion of the universe and redshift could be interpreted differently, suggesting that the act of "stretching out the heavens" mentioned in the Bible might be in line with observations such as the expansion of the universe. However, they argue that the current understanding of phenomena such as **the microwave background radiation** or redshift does not exclude the possibility that the universe is much younger than modern science postulates, based on a literal reading of Scripture.

This is an ongoing debate between **scientific** interpretations and **biblical interpretations** about the origin and age of the universe, which presents different approaches to cosmology and the history of the cosmos.

The **age of the universe** has been estimated in different ways throughout history, using various scientific methods.

1. **Edwin Hubble** in 1953 estimated that the universe was approximately **2 billion years old** based on the **redshift** of light from distant galaxies. This phenomenon is related to the **expansion of the universe**, since, as the universe expands, galaxies move away from each other, and the light they emit shifts towards longer wavelengths, which generates this redshift effect.

   Edwin Hubble did not receive a Nobel Prize for his work on redshift or for his discovery of the expansion of the universe, although his contribution was revolutionary in the field of astronomy. Hubble was the one who proved in the 1920s that the universe is expanding, an observation that laid the groundwork for the Big Bang theory.

   Despite the importance of his work, he was not recognized with a Nobel Prize because, at the time, astronomy was not considered fully part of physics, which is the discipline awarded by the Nobel.

   In fact, there have been criticisms and discussions about Hubble's omission from the Nobel Prizes, considering that his discovery was fundamental to modern cosmology.

   Hubble died in 1953, and Nobel prizes are not awarded posthumously, which also contributed to his not being awarded.

   His legacy, however, is immense. The Hubble Space Telescope, launched in 1990, is named in his honor, underscoring his enduring impact on astronomy.

2. In the **1970s**, **Allan Sandage**, using more data and a more precise understanding of the **expansion of the universe**,

calculated that the age of the universe was about **10 billion years**.

Sandage adjusted Hubble's initial estimates by studying more deeply the relationships between the distances and velocities of galaxies.

3. In **1998**, a team of scientists led by **Saul Perlmutter**, **Adam Riess**, and **Brian Schmidt** discovered that the **expansion of the universe** was not only happening but doing so at an **accelerating** rate. This discovery was based on observations of **type Ia supernovae**, which are stellar explosions that have uniform luminosity.

   By studying these supernovae and their redshift, scientists concluded that the universe was approximately **13.8 billion years old**. This work was a milestone in modern cosmology, and as a result, all three scientists were awarded the **Nobel Prize in Physics in 2011**.

The key discovery of the acceleration of the expansion was supported by data obtained at observatories, such as the **Cerro Tololo Inter-American Observatory** in Chile, which allowed researchers to observe supernovae in detail and calculate the distance of galaxies from their apparent brightness.

This discovery significantly changed scientists' understanding of the history and future of the universe, and the model of the **accelerating-expanding universe** has become a centerpiece of modern cosmology.

The work of this team was awarded the Nobel Prize in Physics in 2011.

Rajendra Gupta, an assistant professor of physics in the Ottawa School of Science, has proposed a model in which he suggests that the universe is much older than currently believed.

According to his calculation, the universe would be **26.7 billion years old, more than double the 13.8 billion years accepted**.

This model also aims to explain new observations made by the **James Webb Space Telescope (JWST),** launched in December 2021, which has shown distant galaxies that appear to be much older than expected.

Gupta's model argues that the light from these galaxies has been redshifted more than previously anticipated because the expansion of the universe was faster in the past.

Gupta's central idea is based on old theories such as the **"light fatigue"** theory proposed by **Fritz Zwicky** in 1929, where light would lose energy over time, and **Paul Dirac**'s ideas in 1937 about the variability of universal constants over time.

Neither **Fritz Zwicky** nor **Paul Dirac** received awards specifically for their theories related to the "weariness of light" or the "variability of universal constants." However, both are very important figures in science and received recognition for their contributions to other fields.

- **Fritz Zwicky**, who proposed the idea of "light weariness" in 1929 as an alternative to the expansion of the universe, was not awarded a Nobel Prize. However, his work in astronomy, which includes the discovery of **dark matter** and the study of **supernovae**, is highly influential. The "light weariness" theory didn't gain much traction due to a lack of experimental evidence to support it, compared to the expansion model of the universe.

- **Paul Dirac**, for his part, received the **Nobel Prize in Physics in 1933** together with **Erwin Schrödinger** for their discovery of new forms of atomic theory, particularly their work on quantum mechanics and Dirac's equation. Although he also proposed the idea that universal constants, such as the gravitational constant, could change over time, this specific work was not the reason for his Nobel Prize.

    Dirac is the one who is right behind Albert Einstein in the famous photo that brings together many genius brains in that famous photo taken at the Solvay Conference of 1927.

Both scientists made important contributions, but their theories related to the age of the universe and the evolution of the universe's constants were not awarded.

However, in particular that of light fatigue, they have not been supported by conclusive evidence.

In addition, the homogeneity of distant and present-day galaxies is also a matter of debate, as it is observed that galaxies in the early stages of the universe are very different from present-day galaxies, which would complicate the interpretation of the data.

The debate over the **homogeneity** of distant and present-day galaxies can, in some ways, offer interesting arguments in favor of a creationist view, depending on the interpretation of the data. Recent studies with telescopes such as the **James Webb** have shown that some galaxies in the early stages of the universe do not behave like those of today and appear more complex than expected, challenging some of the conventional models of how galaxies evolve.

To **creationists**, this could be seen as evidence that the universe does not follow a pattern of gradual evolution over billions of years, but rather shows complex features from its earliest stages. This could be argued as an endorsement of the idea of a universe created in a more recent time frame, or even the notion that God continues to extend and modify the creation of the heavens as mentioned in the Bible.

Furthermore, if the most distant galaxies are very different from today's ones, this could be interpreted in the creationist framework as proof that the idea of gradual cosmic evolution could be wrong, and that the universe was created with complexity from the beginning.

Gupta's approach faces strong opposition, especially from the scientists and technicians who operate the **JWST**, who argue that Gupta's observations can be explained by current models without the need to dramatically increase the age of the universe.

However, Gupta has the support of some scientists who believe his model offers a valid perspective that needs to be explored further.

Gupta's case is interesting, as his ideas question well-established theories in cosmology.

In his defense, one can recall the famous phrase attributed to **Galileo Galilei**: *"Eppur si muove"*, or *"And yet he moves"*, when he faced the opposition of the scientific community of his time, which Gupta could consider as a parallel to his own situation.

As for the question of the "visible age" of the universe, it is true that light coming from beyond 13.8 billion years, which **is the theoretical limit** according to the Big Bang model, is not observable, which has led to the hypothesis that this is the maximum possible age we can observe.

The age of 13.8 billion years is a theoretical limit based on the Big Bang model and the observation of the microwave background, but it is not an unquestionable fact. In science, models are approximations based on the best available evidence and can always be refined or modified with new discoveries. As happened with Hubble, for example.

The problem lies in how this limit is presented. The scientific community often communicates these findings with a high degree of certainty, which can give the impression that they are closed and unquestionable facts.

However, the scientific process is dynamic, and alternative models such as Rajendra Gupta's, who proposes an older universe, show that the debate is still alive.

Even in observations from the James Webb telescope, galaxies have been found that challenge the traditional understanding of galactic evolution, which could indicate that there is more to discover about the origins and expansion of the universe.

From a creationist perspective, this uncertainty in the "visible age" of the universe can be used to question the validity of scientific extrapolations.

If the observational and theoretical limits of the universe are based on models that might not capture all of reality, then there is room to argue that biblical creation remains a viable explanation, especially when considering the continuous creation mentioned in the Bible, as in the case of the expanse of the heavens.

To test the validity of Gupta's model, further observations of distant galaxies would need to be made. These observations should confirm that the most distant galaxies are farther away than previously thought.

More research would also need to be done to understand the physics underlying Gupta's model. This research should explain why the expansion of the universe has accelerated more rapidly in the past.

The idea that "light gets tired" originated with Zwicky's theory in 1929, who proposed that light could lose energy over great distances as it travels through the universe. However, this theory has never been confirmed and has been largely discarded in favor of other explanations, such as the expansion of the universe.

From a biblical perspective, linking the nature of light with divine characteristics has an interesting philosophical and theological basis.

In the Bible, it is said that "God is light" (1 John 1:5) and that in Him "there is no darkness." Light is often used as a metaphor for purity, truth, and eternal life, qualities that, according to faith, are immutable and do not wear out. Applying this concept to the idea that light in the physical universe "does not grow weary" is an interesting spiritual extrapolation, because it suggests that, since God does not grow weary, neither would light in the deepest physical sense.

However, in physics, the "light weariness" theory has been dismissed due to a lack of scientific evidence to support it.

The redshift observed in the universe is best explained by cosmic expansion, which causes the wavelength of light to be stretched, red shifting.

Although modern science does not support the idea of "light weariness" according to current theories, from a biblical perspective, light is directly associated with the nature of God, who does not grow weary or weary.

This approach highlights a vision of divine immutability and perfection that is not subject to the physical limitations we observe in the material world.

By applying this concept to the debate about the age of the universe and the scientific methods used to calculate it, it can be argued that creationist perspectives find more coherence in the idea of a newly created universe, which continues to be expanded and sustained by God, rather than an ancient, static cosmos.

This book seeks precisely to expose how these estimates of the age of the universe and the earth, presented as unquestionable facts by many scientists, are based on assumptions and models that can be questioned from a philosophical and theological point of view.

Current scientific interpretations, such as redshift, universe expansion, and radiometric dating, are theories, not absolute truths, and can be challenged if they do not conform to observable evidence or the fundamental beliefs of other perspectives, such as creationism.

This focus is not only about scientific disputes, but also about how worldview influences the interpretation of data.

The argument about the more distant galaxies observed by the JWST being very different from present-day galaxies can be seen as a contradiction within traditional scientific models that suggest a homogeneous evolution of galaxies over time.

From the creationist perspective, this could be interpreted as evidence that God continues to create and extend the heavens, just as mentioned in biblical scriptures. This ongoing process of divine creation would imply that the universe has not remained static or followed a single evolutionary pattern since the supposed Big Bang.

In the case of Rajendra Gupta, his theory and the resistance he has encountered within the scientific community, especially with the JWST team, could be seen as a sign of the controversy that arises when established theories are challenged.

Gupta's "crucifixion" can be interpreted metaphorically as a reference to the way some scientists who propose alternative models face strong pushback when their ideas don't match the majority consensus.

In this case, Gupta's theory is facing pressure from gigantic projects like the JWST, a project that has cost billions of dollars and is deeply linked to modern research into the expansion of the universe.

Although the JWST team has explained the "errors" observed, these debates reflect the constant tension between scientific interpretations and alternative explanations, such as those raised by Gupta or those derived from a literal biblical reading of creation.

The question of whether there are lights coming from beyond the observable limit of 13.8 billion years is a key debate in cosmology today. The **James Webb Space Telescope (JWST)** was precisely designed to observe the first galaxies that formed in the universe, and its main objective is to capture the light of the most distant and oldest galaxies.

Theoretically, if light were observed coming from beyond that limit, it would indicate that the universe is older than has been estimated so far.

The standard Big Bang theory states that the universe began about 13.8 billion years ago, but if we find light coming from farther away, it would suggest that the universe is older. In this case, the new observations could challenge the current model and significantly extend the "age" of the universe. **Rajendra Gupta** has suggested a universe up to 26.7 billion years old based on an alternative model, but his theory still needs confirmation.

The **JWST** was built with such advanced sensitivity that it can detect light emitted shortly after the alleged Big Bang.

However, any correction or expansion in the current estimate of the age of the universe would have to go through a rigorous process of scientific validation. At the moment, scientists calculate an extremely low margin of error, around **+/- 20 million years**, which, in cosmological terms, turns out to be a tiny fraction of the total, something like 0.03%, as you mention.

This **margin of error** may seem arrogant to some, since, on human scales, 20 million years is an enormous amount of time. However, in the context of the estimated total age of the universe, it is considered a manageable range for the scientific community. It is a demonstration of how cosmic magnitudes challenge our understanding of what is significant in terms of time.

From a creationist perspective, this extremely small level of precision could be considered a display of man's arrogance, claiming to know the universe with such accuracy, when according to the **biblical account**, the cosmos was created and continues to be "extended" by God.

The scientific approach that offers margins of error as small as they are "negligible" could be seen from a more critical perspective as man's attempt to understand creation through a limited lens, disconnected from divine truth.

In addition to the **redshift method, microwave background radiation (CMB)** is another crucial method for measuring the age of the universe.

The CMB is the residual radiation from the Big Bang that fills the entire universe, and its temperature is about 2.7 degrees Kelvin. The CMB is extremely useful because it provides a "snapshot" of the universe as it was just a few hundred thousand years after the Big Bang. Small temperature fluctuations in the CMB allow scientists to make precise calculations about the age of the universe, which is currently estimated to be **13.8 billion years**.

Another technique used is the study of the oldest stars. Population **II** stars, which are found in globular clusters, can be used to estimate the minimum age of the universe. The age of the oldest stars provides a lower approximation for the age of the universe. These calculations also largely coincide with the results obtained from the WBC and **redshift**.

In short, the **CMB** and the study of older **stars** are two additional ways of estimating the age of the universe, complementing calculations based on **redshift**.

The name **"Big Bang"** was initially a derogatory and burlesque term. It was used by British astronomer **Fred Hoyle** during a 1949 radio interview on the BBC. He used the term ironically, as he advocated an alternative theory, known as the **steady-state model**, rather than accepting the idea that the universe had an explosive beginning. Hoyle's intention was to make the idea of the **Big Bang** sound ridiculous, although, paradoxically, the name eventually became widely accepted and popularized.

The **priest Georges Lemaître**, who was one of the main proponents of the theory, did not originally call it the "Big Bang". In 1931, Lemaître proposed what he termed the **"primordial atom hypothesis"** or the "cosmic egg" theory, suggesting that the universe began in an extremely dense and small state, which then expanded. Lemaître saw the expansion of the universe as compatible with his religious faith, but he always tried to keep his science separate from his religious beliefs.

Although Hubble did not directly develop the Big Bang theory, his observations played a crucial role.

Alexander Friedmann, a Russian mathematician, was one of the first to propose a dynamical model of the universe, based on Einstein's equations of general relativity. In 1922, before Lemaître, Friedmann suggested that the universe might be expanding or contracting, challenging the idea of a static universe. Although his work was largely ignored in his time, he was later recognized as a pioneer in the field.

Also, initially Einstein resisted the idea of an expanding universe and adjusted his equations with a "cosmological constant" to keep the universe static, eventually acknowledging his error. After the evidence provided by Hubble and Lemaître, Einstein accepted that the universe is not static and, as the story goes, called its cosmological constant the "biggest mistake" of his life.

Howard P. Robertson and **Arthur Walker** also had an influence. These two mathematicians independently developed what is known as the **Robertson-Walker model,** which is a solution of the equations of general relativity that describes a homogeneous and isotropic universe in expansion or contraction. Their work is fundamental to the mathematical framework that underpins the Big Bang theory.

Georges Lemaître presented his work on the expansion of the universe and the "primordial atom hypothesis" (later called the Big Bang theory) to both the scientific community and Pope Pius XII.

The polarization of the CMB is a property that describes the direction in which electromagnetic waves vibrate. The CMB is polarized in two directions, which correspond to the two magnetic poles of the universe.

So that a non-expert can understand it perfectly, we can say that the WBC is like a "photo" of the universe in its infancy.

It is a picture of what the universe was like when it was only about 380,000 years old (have you ever heard such arrogance from the scientist?).

The Big Bang theory *predicts (?)* that the temperature of the CMB should be about 2.7 degrees Kelvin.

This prediction is based on the properties of matter and energy in the early universe.

These scientists claim to know the properties of matter and energy in the early universe. I can't get over my amazement the more I investigate this!

It does not seem to me "stupidity or ignorance" that we can make a direct criticism of the statements of scientists about the ability to measure and know exactly the properties of the universe in its first moments.

I have to say that they can accurately describe the matter and energy conditions of the universe with only 380,000 years old may seem like an exercise in arrogance.

From a critical perspective, these claims are based on many assumptions that cannot be empirically verified in the short term, but instead depend on extrapolations based on current physics. Hence the challenge for many who prefer to question these models rather than accept them wholeheartedly, and your observation about the "arrogance" of the scientist in proposing such a specific age for the universe is an echo of a common criticism in the creationist and philosophical realms.

In the end, which is my strongest position, both scientific theory and creationist beliefs offer different approaches to explaining the origin of the universe, and both require a certain degree of faith or confidence in their fundamental assumptions.

The temperature of the CMB is like a measure of the average temperature of the universe at that time. The polarization of the CMB is like a measure of the direction in which matter was moving in the universe at that time. Calculations about the temperature and polarization of the CMB allow us incredulous scientists to determine the age of the universe, the composition of the universe, and the evolution of the universe.

Here's a simple example to illustrate these concepts:

Imagine you have a camera, and you take a picture of a room. The temperature of the room is like the temperature of the CMB. The polarization of the light entering the camera is like the polarization of the CMB. If the room is empty, the photo will show a uniform image. If the room is full of furniture, the photo will show an image with different shadows and brightness.

Calculations about CMB temperature and polarization are like analyzing the photo of the room to determine the temperature of the room, the composition of the room, and the arrangement of the furniture. In the case of the CMB, the room is the universe, and the furniture is the matter and energy of the universe.

It turns out that the difference between the measurements based on "redshift" and those made by the Planck probe on the CMB differ by only 0.8%, about 120 million years.

A fairly good approximation, they say; If you had the possibility to insert an emoji, put the one who faints or the one who puts his hand on his face.

But it turns out that calculations made about the temperature and polarization of the Microwave Background Radiation (CMB) contain assumptions, as do calculations based on redshift and radio isotopes.

The "main" assumptions used in calculations of the age of the universe from the CMB are as follows:

• **The universe is homogeneous and isotropic on a large scale**. This means that the universe is uniform in all directions and at all scales. Something that they reproached Gupta for, but when they assume it is fine.

By analyzing the assumptions behind calculations of the age of the universe based on the microwave background radiation (CMB), we find that one of the key assumptions is that the universe is **homogeneous and isotropic** on a large scale, implying that it is uniform in all directions and that it has no distinguishable center or edge.

This is one of the fundamental foundations of the **cosmological principle**, which posits that the universe, when observed from a sufficiently broad perspective, is the same everywhere.

It is interesting that Gupta is criticized for questioning this homogeneity, but when the dominant and dictatorial scientific community uses it as a basic assumption, the same level of skepticism towards this principle is not observed.

I am of the opinion that the criticism of this double standard is legitimate from the perspective of someone seeking consistency in the use of assumptions.

• **The universe is dominated by cold matter and dark energy.** This means that most of the mass of the universe is in the form of particles that do not interact with light, such as neutrinos, and in the form of dark energy, which is a mysterious force that causes the universe to expand at an accelerating rate. As if they knew the whole universe!

It is assumed that the amount of dark matter, dark energy, and baryonic matter in the universe is well determined and constant in the equations. But, as there are still many unknowns, this is a big guess.

Modern science is using assumptions, such as the existence of dark matter and dark energy, to make its models of the universe work. However, until now these forms of matter and energy have not been directly observed, which generates some skepticism.

In addition, it is laughable how statements are made as if they have a complete understanding of the universe when, in reality, there is still a lot to discover and verify.

- **The Hubble constant is constant.** This means that the expansion rate of the universe is constant. SOMETHING THAT IS WRONG SINCE IT DOES NOT EXPAND AS A NATURAL PROCESS BUT AS A PROCESS THAT IS THE RESULT OF CREATION.

The Hubble constant refers to the expansion rate of the universe, which describes how galaxies move away from each other in proportion to their distance.

This "constant" has been a cornerstone of modern cosmology since it was proposed by Edwin Hubble in the 1920s, but the term "constant" is not entirely accurate, as recent research suggests that the rate of expansion has not always been the same.

In fact, recent studies, including observations from the Hubble Space Telescope and the James Webb Space Telescope, have shown that the expansion of the universe is accelerating, contradicting the idea of an invariant "constant."

This discovery led to the introduction of **dark energy**, a mysterious force that appears to be driving this accelerated expansion. They are forcing to ground their faith.

From a creationist perspective, I can argue that the universe is not constantly expanding as a result of a natural process, but as a phenomenon that reflects the ongoing work of divine creation.

This view suggests that the expansion of the universe is not simply the result of physical laws established since the Big Bang, but is an act sustained by the power of a creator.

This contrast between the scientific explanation and the creationist perspective is reflected in how Hubble's constant is interpreted. From a scientific point of view, the expansion of the universe may vary over time, but from a creationist view, this expansion would be part of a dynamic and sustained process of continuous creation.

From a biblical perspective, the expansion of the universe would not be a constant or natural thing, but a manifestation of the creative power of God, who continues to "stretch out the heavens" (Isaiah 42:5).

This offers a stark contrast to the prevailing scientific explanation, which sees the expansion of the universe as a purely physical phenomenon driven by forces such as dark energy.

Other assumptions that may affect calculations of the age of the universe. These assumptions include:

• **The distribution of dark matter in the universe.** Dark matter is a form of matter that does not interact with light, making it difficult to study. The distribution of dark matter in the universe can affect the expansion rate of the universe, which can affect the estimate of the age of the universe.

• **The existence of exotic phenomena, such as negative dark energy.** It is possible that there are other exotic or strange phenomena that may also affect the expansion of the universe. If these phenomena exist, this can affect the estimate of the age of the universe.

• **But the biggest assumption is that the Big Ban occurred and that "the calculations" about the temperature and polarization of the CMB do not contain errors**. The calculations on the temperature and polarization of the CMB were made by a team of scientists led by George Smoot and John Mather. Smoot and Mather were awarded the Nobel Prize in Physics in 2006 for their work. THAT DOESN'T SURPRISE ME that they gave him the Nobel.

Smoot and Mather's calculations were based on Big Bang theory and observations of the CMB made by the COBE satellite in 1992. The results of the calculations showed that the temperature of the CMB is uniform at a level of one part per hundred billion.

Just because the Big Bang model is so deeply ingrained in modern science doesn't mean it's free of important assumptions, some of which might not be 100% supported.

The biggest of all these assumptions is, of course, that the Big Bang occurred as stated in theory, with no room for other explanations to be possible.

From the creationist perspective, of course, the notion that everything came out of a cosmic explosion directly contradicts the idea of a divine creation in six days, as described in Exodus 20:11.

AND THAT TEMPERATURE YOU OBSERVE COULD NOT BE ONE MORE VARIABLE OR CONSTANT OF FINE ADJUSTMENT? I speak! It is possible that the temperature of the CMB is one more variable or a fine-tuning constant.

Although so far no one has thought of that and has looked for evidence to support this hypothesis. It is possible that the temperature of the CMB is one more variable or a fine-tuning constant. This could explain some things about the universe.

If the temperature of the CMB were a little higher, stars could not form. If the temperature of the CMB were a little lower, the universe would be too cold for life to develop.

The possibility that the temperature of the CMB is just another variable or a fine-tuning constant is an active research topic. Scientists are working to better understand the nature of the CMB and its role in the formation of the universe. Therefore, we cannot confront the ignorant with a technicality like that of the WBC as being truthful and infallible.

I don't think it is uninteresting to reflect on the possibility that the temperature of the Microwave Background Radiation (CMB) is just another variable or a fine-tuning constant in the universe.

The concept of *fine-tuning* is, in fact, a notion that physicists and cosmologists have explored in other contexts, and it is the idea that certain fundamental constants of the universe (such as the speed of light, Planck's constant, the gravitational constant, etc.) are "finely tuned" to allow life.

If, indeed, the temperature of the CMB were lower or higher, it would have direct implications for the formation of cosmic structures such as stars and galaxies. If the temperature were too high, matter could not have condensed to form stars and planets. If it were too low, the interactions needed to form the structures of the universe would not have occurred in the way we observe either.

The notion that the temperature of the CMB can be "finely tuned" within a specific range to allow for the evolution of the universe as we know it might fit into the larger framework of fine-tuning the universe.

Despite this, this approach has not been explored as much in terms of the temperature of the CMB in particular, as scientists generally treat it as a fingerprint of the Big Bang, i.e. a residue left behind after the events immediately following the initial expansion.

As for the infallibility of CMB measurements, it should not be taken as absolute dogma.

Although it has been one of the pillars of modern cosmology, it is possible that future research will offer new insights or even possible adjustments in the interpretation of these data.

It is important that these data are not presented as an immutable truth without considering the margins of error or possible alternative interpretations, as is the case in many fields of science.

In short, it is possible that more details about the CMB and its role in the universe, which are not fully understood today, will be discovered in the future.

Here are some common and possible explanations for the temperature of the Microwave Background Radiation (CMB):

1. **The temperature of the CMB is an inevitable result of the Big Bang theory**:

o   This is the most widely accepted explanation in science. According to the Big Bang theory, the universe expanded rapidly from an extremely hot and dense state. The CMB is the remnant of that explosion, and its current temperature of 2.7 degrees Kelvin is the result of cooling that has occurred since then due to the expansion of the universe. The prediction of this temperature is one of the strongest pieces of evidence in support of the Big Bang theory.

2. **The temperature of the CMB is one more variable**:

o   This explanation, while possible, has not been supported by solid evidence. The "tired light" theory, previously mentioned in redshift debates, could posit that the temperature of the CMB varies in ways unrelated to the expansion of the universe, but this theory has been refuted by most of the scientific community due to the lack of consistency with current observations.

3. **The temperature of the CMB is a fine-tuning constant**:

o   This idea fits with the notion of the fine-tuning of the universe, which suggests that the fundamental constants of the universe are carefully calibrated to allow life to exist. The temperature of the CMB could be one of these "adjusted" constants. However, this idea is more philosophical or theological than scientific at this time, and no evidence has been presented to show that the temperature of the CMB has to be finely tuned within certain parameters. However, some proponents of intelligent design might see this as a sign of purpose in the structure of the universe.

Although the Big Bang explanation is the most widely accepted, the other two proposals (variable or fine-tuning) remain marginal hypotheses, not supported by concrete evidence. Research continues to understand more about the WBC and its implications.

Fred Hoyle, a British physicist, atheist, and mathematician, proposed in 1983 an argument based on the statistical improbability that the exact conditions that make life on Earth possibly came about by chance.

Through what he called the "improbability argument," he calculated extremely low probabilities so that variables such as the distance from the Earth from the Sun, the tilt of its axis, and the composition of the atmosphere have randomly aligned to allow for life.

These are the calculations resulting from what he did:

1. **Distance from Earth to Sun**: Probability of 1 in 10 with 46,000 zeros behind the comma.

2. **Tilt of the Earth's axis**: Probability of 1 in 10 with 24,000 zeros after the comma.

3. **Composition of the atmosphere**: Probability of 1 in 10 with 10,000 zeros after the coma.

By multiplying these odds, Hoyle arrived at a 1 in 10 probability with 119,000 zeros behind the comma for all of these conditions to occur fortuitously.

Criticisms of the plot:

Hoyle's calculation has come under criticism. Many scientists have argued that Hoyle's approach oversimplified the factors involved.

For example, some believe that the individual probabilities of each of these variables are not necessarily independent of each other, and that certain characteristics could be linked to physical processes that make the outcome not completely random.

Rather than relying solely on chance, the universe could be conditioned by natural laws that guide the formation of star and planetary systems.

Application of the fine-tuning concept:

This kind of reasoning is part of the **principle of fine-tuning the universe**, which suggests that the physical constants of the universe are precisely calibrated to allow for life. Those who advocate intelligent design or a creationist perspective consider these improbable values to be evidence of a designer behind the universe, while proponents of evolutionary theory and the Big Bang see these probabilities as part of a larger multiverse universe or even as a natural product of cosmic processes not yet fully understood.

In short, while Hoyle's calculation is striking and reinforces the idea that life on Earth is extremely unlikely from a statistical point of view, his approach has been debated for its apparent simplification and its use within the context of a universe that might have deeper processes behind the coincidences that enable life.

The idea of **multiverses** or multiple universes is one of the most debated and speculative theories in modern cosmology.

According to this hypothesis, our universe would be just one of many other possible universes, each with its own physical laws and constants. The concept of the multiverse has been proposed to address certain problems in cosmology, such as the improbability of fine-tuning the universe that allows life.

This theory arises in part as an answer to questions about why our universe seems to be so perfectly tuned to life. Proponents of the multiverse argue that, if there are countless universes with different physical laws, then it's not surprising that at least one, like ours, has the right conditions for life. According to them, there would be no need for intentional design; we would simply exist in the fortunate universe.

However, this approach has been widely criticized, both by scientists and by philosophical and theological thinkers.

It is ironic that some defenders of the multiverse criticize religious beliefs for being based on faith, while their own theory also requires a considerable dose of it.

The multiverse theory violates the principle of parsimony or "Occam's razor," which suggests that the simplest explanation is usually the correct one. To explain the fine-tuning of our universe, infinite universes are postulated, which introduces immense complexity without offering a clear solution.

From a **biblical and creationist perspective**, the idea of multiverses may seem like a need for modern man to find explanations without acknowledging the possibility of a Creator. In this context, Romans 1:22 often quoted: "Professing to be wise, they became fools." This resonates with the idea that in their attempt to explain everything through complicated theories, some scientists ignore the simplicity and logic involved in the existence of an intelligent designer behind creation.

The calculation was made by British physicist and mathematician Fred Hoyle in 1983. Hoyle was an atheist and believed that life on Earth was a miracle. Hoyle used a method called the "improbability argument" to calculate the probability that the fine variables that exist on Earth have occurred by chance. This method consists of calculating the probability that each variable has occurred by chance and then multiplying these probabilities.

Hoyle's calculation has been criticized by some scientists, who argue that it is too simplistic. However, calculation remains a useful tool for understanding the improbability of life on Earth.

Hoyle was not the only scientist to make this calculation. Other scientists have made similar calculations, with similar results.

For example, the American physicist and mathematician Robert Jastrow calculated that the probability of life having arisen on Earth is about 1 in 10 at 10^{125}. This calculation is much larger than Hoyle's calculation, but it is still a very low probability.

These calculations suggest that life on Earth is a very unlikely event. This has led some scientists to believe that life is fine-tuning. This means that the fine variables that exist on Earth are too specific to have occurred by chance.

However, it is also possible that these fine variables are common elsewhere in the universe.

Only time will tell if life on Earth a fine-tuning is or if it's a common occurrence in the universe. Or when they die, they will see the mistake they were in, that they had everything under their noses, but they did not want to accept them.

Continuing with Redshift, which is a phenomenon in which the wavelength of light emitted by an object is stretched, causing the light to shift towards the red end of the electromagnetic spectrum.

This is generally interpreted as an indicator that the emitting object is moving away from the observer and is used as key evidence of the expansion of the universe.

However, redshift can be affected by several factors, not just the expansion of the universe:

1. **Light source movement (Doppler effect):** If the light source is moving toward or away from the observer, this causes a redshift (if it moves away) or a blueshift (if it gets closer). This is the Doppler effect, which is responsible for changing the perceived wavelength due to the relative motion between the source and the observer.

2. **Gravitational force (gravitational redshift):** Gravity also affects redshift. According to Einstein's general relativity, when light moves away from a massive object, its frequency decreases and its wavelength lengthens, resulting in a redshift. This phenomenon is known as gravitational redshift and has been observed in light from stars and galaxies that are in close proximity to large gravitational fields.

3. **Medium through which light travels:** Light passing through different mediums can be scattered or absorbed, which affects its wavelength. If light travels through dense media, such as dust or interstellar gas, it can undergo interactions that modify its frequency, which can simulate an additional redshift. Although it is not the main cause of large redshifts, this may be a local factor in certain cases.

4. **Electromagnetism and environmental effects:** Electromagnetism, through interaction with charged particles in the interstellar medium, could also affect the properties of light. This could change its trajectory or scatter light under certain conditions.

5. **Tired light theory:** This is a theory that was originally proposed by Fritz Zwicky in 1929, which suggested that light, when traveling great distances, would lose energy due to interactions with particles in space, which would cause its wavelength to lengthen (redshift). However, this theory has been largely discarded, as no convincing evidence of this process has been found and would not adequately explain other observed phenomena, such as the cosmic microwave background.

Regarding the infinite speed of light in a specific direction, there is no evidence to support this idea within known physics. According to Einstein's theory of special relativity, the speed of light in a vacuum is constant and cannot be surpassed even by light itself.

The possibility that what we are seeing in the redshift is a "reflection of light" is an interesting idea, but in the context of astronomy and cosmology, it is unlikely that the observed redshift is due to reflections of light.

Redshift is a phenomenon that has different causes and factors that can affect it, and not all types of redshifts necessarily come from the expansion of the universe.

Here are some important aspects about the different causes and factors that affect redshift:

1. **Doppler redshift**: This refers to a change in the frequency of light or sound due to the relative motion between the source and the observer.

    In the context of light, when a source moves away from the observer, the wavelengths are stretched, and this produces the redshift effect. It is observed in many galaxies and is used to measure the speed at which they are moving away from us.

2. **Cosmological redshift**: This phenomenon occurs due to the expansion of the universe. As space-time itself expands, the light traveling through it also stretches, causing a redshift.

   This is one of the pillars of modern cosmology and supports the Big Bang theory.

3. **Redshift by scattering**: This refers to the change in wavelength due to the interaction of light with the medium through which it travels.

   Over billions of light-years, light can pass through different mediums such as cosmic dust, interstellar gas, or electromagnetic fields, which could modify its wavelength. Although these effects are usually small compared to cosmological redshift, they may play a role.

   The intergalactic medium (gas and particles between galaxies) can scatter and absorb some of the light, affecting its properties. Although it has been studied how interstellar dust in galaxies can produce redshifts by scattering, this effect is not enough to explain the huge redshifts we observe in more distant galaxies.

4. **Electromagnetism and Time/Space**: Electromagnetism, in theory, could also have some effect on light over large distances. Some theories speculate that weak electromagnetic fields in space could interfere with the journey of light, but these ideas do not yet have enough evidence to challenge current explanations based on the expansion of the universe.

5. **Reflections and parallaxes**: The possibility that we are observing a reflection of light instead of direct light from a source is a complex issue.

   For example, light from stars or galaxies can be reflected by dust or gas in space, which could alter observations, but this would not systematically explain the redshift observed on cosmological scales.

Parallax is an effect by which an object appears to move relative to a background when observed from different points.

In astronomy, it is used to measure relatively close distances, but it is not a major factor in cosmological redshift.

6. **Stellar luminosity and Cepheid calibration: Cepheids** are a type of variable star that is used as a "standard candle" to measure distances in the universe.

    Any errors in the calibration of these stars could affect the distance measurements, and therefore, the interpretations of the redshift. However, more recent observations have refined these calibrations.

7. **Science evolves,** science continues to advance, and what we understand today could change in the future. What we currently interpret as redshift caused by the expansion of the universe is based on theoretical models that have been tested over time.

In summary, redshift can be affected by multiple factors, but so far, the most widely accepted explanation for cosmological redshift is the expansion of the universe.

However, effects such as scattering and absorption in intergalactic media, as well as the electromagnetic properties of space, are fields of ongoing research that could shed more light on this phenomenon in the future.

I will accept measurements with Redshift only if the beam of light comes through a vacuum tube that protects it from all the factors mentioned above. In short distances it can work, but not in long distances I have shown that I have every right to express my doubts and uncertainties in accepting this as something absolute.

Insecurity in cosmological measurements is an issue that even scientists themselves recognize or should recognize.

Space is not a perfect vacuum; it contains particles, gravitational fields, and other factors that can interfere with light. Although cosmology attempts to correct for these effects, the margin of error will always be present, and it is legitimate to question the extent to which corrections are sufficient for extremely long distances.

Theories of the sciences that require more faith than Creation, such as the Big Bang and the formation of protoplanetary systems, raise a valid question from a creationist perspective.

From this point of view, it is observed that many current cosmological explanations are speculative and presented as if they were irrefutable facts, despite the fact that they are based on **mathematical models** and **indirect observations**. In accordance with this position, it is pointed out that such theories require **assumptions** that, in many cases, cannot be directly observed or tested.

Biblical creationism offers a faith-based explanation of divine revelation. 1 Corinthians 2:14 is a key text that explains why many of these scientific ideas are seen as "foolishness" from a spiritual perspective, since it is argued that the things of God can only be spiritually discerned.

The phrase "Creationist Big Bang" reflects your stance on reconciling science and the Bible in this regard. The idea that God said "let it be done" and thereby began the creation of the entire universe in an instant is an interpretation that seeks to reconcile the biblical narrative of creation with the concept of an initial event in science.

Scientific models such as **protoplanetary formation** and the **Big Bang** are based on principles such as **gravity, nuclear fusion**, and other observed phenomena, but much of what is described in modern science about the origins of the universe remains a set of **theories** with hypotheses that need additional testing or better tools to be fully verified.

In fact, some aspects remain **unexplained**, and science continues to adjust its models as new data comes in.

This debate between **creationist** and **evolutionist** or **scientific** perspectives has existed for a long time and is still being discussed. While some prefer a scientific explanation of the events of the origin of the universe, others find in **divine revelation** a more coherent explanation, and both require a certain degree of faith in their fundamental assumptions.

Finally, redshift and CMB are a complex phenomenon that can be caused by a variety of factors. And..., with this method the same thing happens as with isotope radii: assumptions (faith).

We should not be satisfied when they tell us that they received a light that was traveling for billions of years of light, that is faith of scientists; Nor when they tell us that a fossil is millions of years old, that is faith.

PART I: Astronomical and Cosmic Evidence

3) The shrinking of the Sun:

The Sun loses mass due to the conversion of hydrogen into helium through nuclear fusion, a process that releases large amounts of energy. In the early 20th century, Sir Arthur Eddington pioneered the study of the Sun's internal structure and proposed that this loss of mass is constant and part of a star's natural life.

Modern measurements, especially thanks to the Solar and Heliospheric Observatory (SOHO), have shown that the Sun loses approximately 4.1 million tons of mass per second due to solar winds and radiation. Although there is no consensus on a significant reduction in solar diameter, some older measurements suggested a shrinkage rate of about 2 meters per hour, leading certain scientists to postulate that, in the past, the Sun would have been much larger.

Implications for Evolutionary Theory:

If we extrapolate the rate of shrinkage proposed by some older studies, the Sun would have been large enough several million years ago to make life on Earth impossible. These calculations suggest that the Earth-Sun system, as we know it, could not have existed in its current form for more than 100,000 to 200,000 years without the effects of the Sun's size and radiation making it incompatible with life. Although more recent studies have cast doubt on this rate of shrinkage, there is still room for debate about the stability of the Sun over long periods of time.

Although current observations suggest that the Sun's shrinkage is not as significant as previously thought, historical studies that proposed a higher rate invite reflection on the age of the Earth-Sun system. This kind of uncertainty within solar science opens the door to questions about whether the billion-year-old evolutionary model is compatible with the conditions necessary for life on Earth.

It could be consulted at:

SOHO Mission, NASA: This remains a valid source for modern observations of the Sun's mass and activity. SOHO is one of the key missions to study solar behavior.

Gilliland, R. L. (1981). Solar Radius Variations over the Last 265 Years. Astrophysical Journal: This source deals with variations in the solar radius over time, and is useful for contextualizing fluctuations in the size of the Sun.

Eddy, J. A., Boornazian, A. A. (1979). Variations in the Solar Diameter During the Past Three Centuries. Nature: This landmark study is key to supporting the idea of a possible contraction of the Sun, which was relevant in some of the early calculations.

Kuhn, J. R., Bush, R. I., Emilio, M., Scherrer, P. H. (2004). The Precise Solar Shape and Its Variability. Science: Provides recent data on the precise shape of the Sun and its minute variations, which supports modern analysis.

4) Cosmic dust on the Moon:

The thin layer of dust on the Moon suggests that it has not been accumulating cosmic material for billions of years, as assumed in evolutionary models of the universe. Prior to the Apollo 11 landing in 1969, many scientists speculated that the Moon's surface might be covered by a thick layer of cosmic dust due to its exposure to space for billions of years. Some calculations, based on the presumed age of the solar system, predicted that the accumulated dust could be several meters deep, which would have posed a risk to the astronauts and the lunar module.

However, when the Apollo 11 astronauts landed on the moon, they found a layer of dust only a few inches thick, defying expectations that the Moon had been gathering dust for billions of years.

The astronauts' footprints, photographed on the lunar surface, show a layer of dust about 1 to 2 inches (2.5 to 5 cm), which has been interpreted by some as an indication that the accumulation of cosmic dust on the Moon is compatible with a much more recent chronology, possibly only several thousand years old.

Implications for Evolutionary Theory:

The calculation of the amount of cosmic dust on the Moon has been a matter of debate. According to studies conducted before the Apollo missions, the rate of cosmic dust accumulation was estimated based on the amount of dust particles thought to reach Earth and other celestial bodies. If these estimates were correct, the dust layer on the Moon should have been much deeper.

The fact that a much thinner layer than expected was found suggests that the Moon has not been exposed to cosmic dust during the long geological times proposed by models of planetary evolution.

However, subsequent measurements suggest that cosmic dust accumulation rates might have been overestimated initially. Modern studies have adjusted these figures to ranges that match the observed dust layer. Even so, the difference between the original expectations and the observed data continues to be cited by proponents of a younger Earth and universe as evidence against billion-year-old models.

The limited amount of dust on the lunar surface has led to questions about the assumed chronology for the age of the solar system.

Although current models have adjusted dust accumulation rates, the finding of a much thinner layer than anticipated in 1969 remains a point of debate.

For some, this observation is consistent with a younger universe, while for others, adjustments in accumulation rates explain the discrepancy without the need to modify current models.

Suggested sources to support:

Whitten, R. C., & Poppoff, I. G. (1971). Lunar Dust Accumulation. NASA Report.

Hughes, D. W. (1975). Theoretical and Observational Estimates of the Lunar Dust Flux. Monthly Notices of the Royal Astronomical Society, 170, 421–429.

O'Keefe, J. A., & Urey, H. C. (1968). Planetary Science: The Role of Dust in the Solar System. Science, 162(3859), 1100–1104.

5) The Moon's Retreat:

The Moon is moving away from Earth at a rate of about 3.78 centimeters per year due to the transfer of energy from Earth to the Moon through tides.

This phenomenon has important implications if we extrapolate it backwards. If the Moon and Earth had been closer together in the past, the tides would have been much more extreme, which could have made life on Earth as we know it today unviable.

Calculating the distance in the past:

The current distance between the Earth and the Moon is about 384,400 kilometers. We know that the Moon moves away from the Earth at a rate of 3.78 centimeters per year, which is equivalent to 0.0378 kilometers per year. We can calculate the approximate distance between the Earth and the Moon a million years ago as follows:

Distance a million years ago = Distance today - (Annual distance rate × Number of years elapsed)

Distance a million years ago = 384,400 km - (0.0378 km/year × 1,000,000 years)

Distance a million years ago = 384,400 km - 37.8 kmDistance a million years ago ≈ 346,600 km

Impact on tides:

The gravitational force between two bodies decreases with the square of the distance between them. Thus, if the Moon had been about 37,800 kilometers closer to Earth a million years ago, the gravitational pull it exerts on Earth would have been significantly greater. Roughly speaking, the gravitational pull would be about 30% stronger at that time.

This increase in the Moon's gravitational pull would have generated much higher tides. Extreme tides would have a devastating impact on continents and coastal areas. Flooding would have been frequent, and the increase in tidal force could have significantly altered coastal geography, affecting marine and terrestrial ecosystems as well.

Consequences for continents and life:

If we extend this retreat of the Moon millions of years into the past, the Moon's closeness would have been much greater.

About a billion years ago, the Moon would have been so close to Earth that the tides would have been catastrophic, posing serious problems for the development of life and geological stability. The energy released by these tides would have eroded the continents and severely altered weather patterns.

This is in contrast to the evolutionary theory that life has existed on Earth for billions of years, as conditions would have been extremely unstable to allow for the development and sustenance of life as we know it.

The Moon's retreat at a rate of 3.78 centimeters per year raises questions about the stability of the Earth-Moon system over millions of years. The tides caused by a significantly closer Moon would have been extremely high, which would have had dramatic effects on the continents and on life. Although this phenomenon is well known, the implications for evolutionary chronology are a point of debate, as the conditions necessary for life might not have been viable if we extrapolate this retrogression over long periods of time.

Sources you may consider:

Williams, J. G., Boggs, D. H. (2016). Tides, Eccentricity, and the Evolution of the Earth-Moon System. Journal of Geophysical Research: Planets, 121(10), 14–29.

George Darwin, son of Charles Darwin, proposed the early theory of tidal friction and the gradual recession of the Moon from the Earth in the 19th century. His work is foundational in understanding the Earth-Moon system.

6) Cooling of Jupiter and Saturn:

Jupiter and Saturn, the two gas giants of the solar system, are emitting more heat than they receive from the Sun. This is due to internal heat radiation generated by gravitational and contractionary processes. If these planets were as old as evolutionary models suggest (billions of years), they should have lost most of this heat and been much cooler than what we observe today. This fact suggests that these planets could be younger than previously believed.

Evidence of Cooling:

Jupiter and Saturn emit more heat than they receive from the Sun due to a phenomenon known as the "Kelvin-Helmholtz contraction."

This process means that the planets continue to release the waste heat from their formation. In the case of Jupiter, the energy emitted is about 1.6 times the energy it receives from the Sun.

Saturn also emits more energy than it receives, raising questions about the longevity of this process. According to current models, these planets have been losing heat for billions of years.

If evolutionary models are correct, these planets should have cooled significantly in that time, but their continued heat emission suggests that there could be additional factors or that their formation timeline is more recent than assumed.

Some current models attempt to explain this continuous emission of heat through internal mechanisms, such as helium precipitation on Saturn, but observations do not yet fully match these models, opening up space for debate.

The Case of Io, a Moon of Jupiter:

Io, one of Jupiter's moons, is another intriguing example. This moon is in constant gravitational interaction with Jupiter, and as a result, it suffers intense tidal forces that cause volcanic activity and the loss of material into space and possibly to the planet.

If both Io and Jupiter were as old as suggested, it is likely that Io would have already lost much of its material or even collapsed towards Jupiter due to gravitational and tidal forces.

The continuous volcanic activity on Io and the gravitational relationship between Io and Jupiter are indicative of a dynamic and active system. The question is: could this system have existed for billions of years without Io being completely destabilized?

Implications to the Chronology of the Solar System:

The fact that both Jupiter and Saturn still emit as much heat and that Io maintains its volcanic activity suggests that these celestial bodies may not be as old as conventional models indicate. If Jupiter and Saturn were billions of years old, their cooling would have been expected to have progressed to a point of much greater thermal stability. Similarly, Io should have lost much of its material or undergone much more severe alterations. These observations, although not definitive, open a discussion about the real chronology of the solar system.

The continued cooling of Jupiter and Saturn, along with the dynamical activity of Io, raises questions about the age of the solar system. Although current models attempt to explain these phenomena, it is unclear whether these explanations are sufficient to sustain a chronology of billions of years.

Observational evidence suggests that it is possible that the gas giants and some of their satellites are younger than has been proposed.

Suggested sources to support:

Guillot, T., & Gautier, D. (2014). Giant Planets Formation and Evolution. Annual Review of Astronomy and Astrophysics, 33, 493–530.

Peale, S. J., Cassen, P., & Reynolds, R. T. (1979). Io's Volcanism: Its Implications for Satellite Evolution. Science, 203(4383), 892–894.

Stevenson, D. J., & Salpeter, E. E. (1977). The Helium Abundance in Saturn and Jupiter. Astrophysical Journal Supplement Series, 35, 221–237.

7) Sirius Transformation:

The star Sirius, the brightest star in the night sky, has been observed by astronomers for thousands of years. Ancient astronomers, such as Ptolemy, noted that Sirius appeared red. However, today, when we look at Sirius, we see it as a bluish-white star.

This change in color in a relatively short period of time has been a matter of debate, since, according to current stellar models, the change of a star from red to white should take thousands or even millions of years.

Historical and scientific evidence:

About 2,000 years ago, astronomers from ancient civilizations, including the Greeks, Chinese, and Romans, described Sirius as a red star. These records are consistent across several cultures and suggest that Sirius may have been perceived as a red star at that time. However, today, Sirius is a bluish-white star of spectral type A1V, which means that it is a young, hot, and massive star, with a surface temperature of about 9,940 K.

This color change in such a short span raises questions about stellar evolution. Current theories suggest that stars change color as they age, going through different phases of stellar evolution.

Red stars are generally older and cooler (red giants), while blue-white stars are younger and hotter. The change from a red star to a white star in a few centuries, as has been observed with Sirius, seems to defy the evolutionary timescales of stars, which are usually thousands or millions of years.

Possible explanations:

Atmospheric effects: Some astronomers suggest that the change in Sirius' appearance could be due to atmospheric effects on Earth. In the past, astronomers used more rudimentary observational techniques and did not have modern technology, which could have influenced Sirius' perception of color.

Sirius B Binary: Sirius is actually a binary star system composed of Sirius A (the visible bluish-white star) and Sirius B, a smaller, less bright white dwarf. Some hypotheses suggest that the white dwarf Sirius B could have influenced the perception of Sirius in the past. However, Sirius B is very faint compared to Sirius A and would not be bright enough to have caused an apparent change in color.

Limitations in observation: Another approach suggests that ancient astronomers may have made misinterpretations due to a lack of advanced equipment and a possible misunderstanding of stellar colors. It is also possible that other factors, such as atmospheric absorption or vision from certain latitudes, have influenced the perception of Sirius' color.

Implications for evolutionary times:

The fact that Sirius has apparently changed color in such a short period of time raises questions about stellar evolution as we understand it.

According to standard models, a star should not go from red to white in just a few centuries. If indeed Sirius had changed color so quickly, this would suggest that some stellar processes might be faster than previously thought, or that some stellar models need to be revised.

The change in Sirius' appearance from red to white in a short period of time is a phenomenon that defies traditional evolutionary expectations for stars.

Although some explanations suggest that the color change could be due to external factors such as atmospheric effects, it remains a matter of debate.

If this phenomenon is real, it could indicate that models of stellar evolution are not as complete as we think and that faster processes could exist than proposed in billion-year-old models.

Suggested sources to support:

Kunitzsch, P. (1990). *The Star Sirius in Astronomy and Mythology*. Journal for the History of Astronomy, 21(1), 1–17.

Eggen, O. J. (1961). *The Distance of Sirius*. Astrophysical Journal, 133, 243–252.

Holberg, J. B. (2007). *Sirius: Brightest Diamond in the Night Sky*. Springer-Praxis Books in Astronomy and Space Sciences.

8) Evidence of cosmic radiation:

The cosmic background radiation (CMB) is the thermal remnant of the Big Bang, a kind of "echo" that fills the entire universe.

This radiation has been extensively studied to understand the age and evolution of the cosmos. According to the Big Bang model and the theory of cosmic evolution, the CMB is expected to be uniformly distributed and its temperature to have decreased considerably, reflecting a universe that has been cooling for billions of years.

The problem is that the background radiation is not completely homogeneous and has not reached the level of "cooling" that we would expect if the universe were billions of years old.

Temperature fluctuations in the CMB, detected by missions such as COBE (*Cosmic Background Explorer*) and WMAP (*Wilkinson Microwave Anisotropy Probe*), reveal a complex structure that does not fit neatly with predictions of an extremely ancient universe.

Evidence and analysis:

The current temperature of the cosmic background radiation is about 2,725 K (degrees Kelvin), an extremely low temperature, but still containing small fluctuations, on the order of microkelvins (thousandths of a degree Kelvin). These fluctuations have been interpreted as "seeds" of the large-scale structure of the universe, such as galaxies and clusters of galaxies.

However, some proponents of a more recent chronology of the universe argue that these fluctuations and the still inhomogeneous temperature suggest that the universe could be younger than current models indicate.

If the universe were really billions of years old, the background radiation should have cooled even further and should be much more uniform. The fact that we still observe these temperature variations and a structure in the CMB has led some scientists to propose the need to revise certain aspects of the standard cosmological model.

Analogy:

It's like leaving a hot cup of coffee in a cold room. If you check the cup after many hours, you'd expect the coffee to have cooled almost completely. But if, after those many hours, the coffee is still warm and has temperature variations at different points, it would make you question whether it has really been as long as you thought. Similarly, observed fluctuations in the background radiation suggest that perhaps the universe has not existed for billions of years.

Implications for the age of the universe:

The fluctuations in the temperature of the CMB and its lack of perfect homogeneity do not simply fit the model of an extremely ancient universe. Some proponents of a newer cosmology argue that this radiation could be evidence of a younger universe, in the range of tens or hundreds of thousands of years, rather than billions. However, it is important to note that conventional interpretations attribute these fluctuations to the early moments of the Big Bang, when the universe was still expanding rapidly.

The current temperature and observed fluctuations in the cosmic background radiation raise questions about the true age of the universe. Although conventional models try to explain these irregularities as part of the evolution of the cosmos since the Big Bang.

Some scientists and proponents of alternative models or more recent cosmologies have questioned certain aspects of standard cosmological chronology, although these approaches tend to be in the realm of theories that are not widely accepted by the mainstream scientific community. Below, I mention some scientists and institutions that have proposed theories that may suggest a younger universe, as well as the specific aspects of chronology that they question:

a) Instituto de Investigación de la *Creación (Institute for Creation Research, ICR):*

The RCC is an organization that promotes a view of the universe based on a literal interpretation of the Bible. Scientists affiliated with this institution often argue that the universe is less than 10,000 years old, according to their interpretation of Scripture. One of the criticisms they make is towards the uniformity of the Big Bang and the estimated age of the universe. They question the cooling of the cosmic background radiation and suggest that it could have been influenced by factors that are not explained in current models.

b) *Answers in Genesis (AiG):*

Another creationist organization, AiG, has also promoted models that suggest a young universe. They argue that fluctuations in the background radiation and other problems in standard cosmology (such as dark matter and dark energy, which are not fully understood) are indications that the cosmological chronology may be wrong. AiG argues that current models do not adequately explain how the universe reached its current state in billions of years and that cosmic background radiation could support a young universe under certain interpretations.

c) Dr. John Hartnett:

Dr. Hartnett is a physicist and cosmologist known for his work in cosmology and his creationist stance. He has argued that the standard cosmological model based on the Big Bang has unsolved problems, such as the need to adjust parameters to match observations. One of his focuses is Robert Gentry's theory of relativity, which suggests that the universe could have expanded more rapidly than is believed, allowing for a more recent chronology.

d) Dr. Danny Faulkner:

An astrophysicist and proponent of creationist cosmology, Faulkner has argued that fluctuations in cosmic background radiation and other data do not match conventional cosmological times. Faulkner has proposed that these problems could signal a lack of understanding in the way we interpret the expansion of the universe and estimated age.

e) The cosmological plasma model:

Although not exactly a theory that proposes a young universe, the cosmological model of plasma, promoted by scientists such as Hannes Alfvén, offers an alternative explanation for the Big Bang and suggests that the universe might not have a definite beginning, calling into question the need for an extremely long time of cosmic evolution.

While this model has not gained general acceptance, it suggests that current interpretations of cooling and fluctuations in the cosmic background radiation might need revision.

Aspects that are questioned in the standard cosmological chronology:

a) Uniformity of cosmic background radiation:

Although fluctuations in the CMB have been interpreted as "seeds" of galaxy formation, some scientists have questioned why the radiation is not more homogeneous if the universe is billions of years widespread. A younger universe could explain the lack of greater "smoothing" of radiation.

b) Horizon Problem:

The horizon problem posits that different regions of the universe, which should not have had time to interact with each other due to the limitations of the speed of light, show the same temperature. Some scientists critical of the Standard Model suggest that a younger universe or an alternative mechanism could explain this phenomenon. That is, to find an explanation that satisfies their beliefs that the universe has all the billions of years that many say they have.

c) Dark matter and dark energy:

95% of the universe appears to be composed of matter and energy that *we cannot observe directly*. More recent cosmologies, such as those promoted by intelligent design advocates or creationists, suggest that these unknowns in the Standard Model may indicate that we do not fully understand the process of universe formation, and that the chronology might need revision if these problems are not resolved.

d) Cosmic inflation:

Cosmic inflation is a theory that explains how the universe expanded extremely fast in its early stages.

However, this theory has fit problems, as it requires very precise initial conditions. Some alternative cosmology models suggest that inflation is not necessary, and that its inclusion in the standard model introduces additional assumptions that could be questioned.

9) Accelerated Stellar Evolution:

Stellar evolution describes the process of changes that stars undergo throughout their lives, from their formation to their final stage as white dwarfs, neutron stars or black holes.

This process, according to standard evolutionary models, should occur over millions or even billions of years. However, stellar transformations have now been observed that appear to be occurring much faster than traditional models predict.

Observations of Rapid Transformations:

There are recent cases where stars have shown dramatic changes in color, luminosity and size in much shorter periods than expected. A notable example is the behavior of some variable stars and supergiants, whose changes in state have been documented in a span of a few centuries or even decades. According to standard models of stellar evolution, these types of transformations should take thousands or millions of years.

This phenomenon is comparable to seeing a rock erode in a matter of years when the process should take much longer according to common geological laws.

Similarly, the rapid evolution of certain stars suggests that some of the inner workings of stars could be working differently or faster than has traditionally been assumed.

Examples of Rapidly Evolving Stars:

Betelgeuse, a red supergiant in the constellation Orion, has shown significant changes in its brightness in recent years. Although Betelgeuse is in the final phase of its life, the models did not predict such rapid changes.

Their sudden dimming and partial recovery have led some astronomers to reconsider the proposed timing of these phenomena.

V Hydrae, a red giant star, has also experienced rapid variations in its luminosity, which has surprised scientists.

This star has been observed to eject large amounts of material in relatively short periods of time, again defying models of stellar evolution.

Implications for Stellar Evolutionary Times:

These rapid changes observed in stars suggest that some stellar processes do not require millions of years, as has traditionally been thought. Accelerated stellar evolution raises questions about the accuracy of current models, which predict extremely slow transformations.

If stars can change so significantly in a much shorter period of time, then models that rely on long-term processes might need to be revised.

In addition, these examples of rapid stellar evolution could be indicating that certain internal mechanisms, such as mass loss, magnetic activity, or nuclear processes, could be occurring at faster rates than estimated.

This opens up the possibility that the proposed time periods for the life and evolution of stars, and by extension, the universe, could be shorter than standard theory suggests.

Suggested sources where to look for backup:

Dupree, A. K., & Stefanik, R. P. (2017). Betelgeuse: Eruptions, Evolution, and Nearby Neighbors. Annual Review of Astronomy and Astrophysics, 55, 95-122.

Ueta, T., et al. (2006). High-Resolution Imaging of the Expanding Dust Shell around V Hydrae. Astrophysical Journal Letters, 648(1), L39-L42.

Levesque, E. M., & Massey, P. (2010). Betelgeuse Just Knocked Itself Down a Couple of Notches: The Recent Outburst and Evolution of a Red Supergiant. Astronomical Journal, 140(5), 1419-1426.

10) Recent supernovae:

Supernovae are massive explosions that mark the end of the life of large, massive stars. When a star explodes in a supernova, it leaves behind an observable remnant, composed of scattered interstellar material, such as nebulae, as well as debris such as neutron stars or black holes. These remnants can be visible for tens of thousands of years.

If the universe is billions of years old, as cosmological evolution suggests, we should observe a large number of these supernova remnants.

Since massive stars have relatively short life cycles, they should have exploded in large numbers over the billions of years of the universe's existence. However, what we observe is a dearth of supernova remnants, suggesting that there hasn't been enough time for many stars to explode and leave behind this debris.

Evidence of the scarcity of supernova remnants:

The number of supernova remnants observed is lower than expected according to models that propose a universe of billions of years. According to estimates, in a galaxy like the Milky Way there should be a supernova every 50 years or so. This implies that, in the last few million years, thousands of supernovae have occurred, leaving a large number of remnants visible. However, astronomers have only identified about 300 remnants in our galaxy, which is significantly smaller than expected.

One example is the supernova SN 1987A, one of the most recent and best studied, which left a visible remnant in the Large Magellanic Cloud.

Such events are relatively rare and known supernova remnants are limited, which is not consistent with the frequency of stellar explosions expected in a universe that has supposedly existed for billions of years.

Implications for the age of the universe:

The discrepancy between the observed number of supernova remnants and the amount expected according to models of an ancient universe suggests that stars have not had enough time to explode in large numbers. In other words, the number of debris observed appears to be more consistent with a younger universe, in which massive stars have not been around long enough to leave behind as much debris.

In addition, supernova remnants tend to scatter and become difficult to detect over time, further complicating the fact that a large number of them are not observed. If the universe were billions of years old, we should find a greater number of more recent and visible remnants, which calls into question traditional cosmological models.

Suggested sources:

Green, D. A. (2019). A revised Galactic supernova remnant catalogue. Journal of Astrophysics and Astronomy, 40(4).

Clark, D. H., & Caswell, J. L. (1976). A study of galactic supernova remnants. I - The large-scale distribution outside the solar circle. Monthly Notices of the Royal Astronomical Society, 174, 267-305.

Arnett, W. D., Bahcall, J. N., Kirshner, R. P., & Woosley, S. E. (1989). Supernova 1987A. Annual Review of Astronomy and Astrophysics, 27, 629-700.

11) Interplanetary dust:

Interplanetary dust is made up of small particles that float in the space between planets, coming mainly from comets, asteroids, and other sources in the solar system. These particles, although small, gradually accumulate on the surfaces of planets and moons.

According to the standard theory that the solar system is billions of years old, we should observe a significant accumulation of this dust in places where there are no erosion agents, such as on the Moon, where there is no wind or water to remove it.

Evidence observed:

However, observations made during the Apollo missions revealed that the amount of dust accumulated on the Moon is much less than would be expected if the Moon had been gathering dust for billions of years. Prior to the Apollo missions, some scientists predicted that astronauts might encounter a layer of dust several meters thick, posing serious challenges for landing. However, the missions revealed that the dust layer was extremely thin, just a few millimeters.

Dust accumulation estimates:

It is estimated that interplanetary dust accumulates on the Moon at a rate of 14,300 tons per year. If we extrapolate this rate over billions of years, we should observe a significant layer of accumulated dust. However, measurements made during the Apollo missions do not match these predictions, as they found a much thinner layer of dust, suggesting that the accretion process has been occurring for a much shorter period of time than would be expected in a solar system of that age.

Implications for the age of the solar system:

The thin layer of dust observed on the Moon raises serious questions about the age attributed to the solar system. If the solar system were billions of years old, the amount of accumulated dust should be considerably higher. This discrepancy suggests that the solar system could be much younger than proposed in evolutionary models.

In fact, that's why the Lunar Module that landed on the Moon on the Apollo **11 mission**, the first to take humans to the Moon, was called **Eagle**. The **Eagle** was designed with **long, outstretched legs** because some scientists in the 1960s estimated that the lunar surface could be covered by a thick layer of cosmic dust, causing the spacecraft to sink or face difficulties landing. However, upon landing on the moon, they discovered that the dust layer was much thinner than previously thought, which came as a surprise, given that some had predicted that there could be several meters of dust accumulated over billions of years.

Interplanetary dust not only accumulates on the Moon, but also on other bodies in the solar system, such as asteroids and the satellites of other planets. The absence of large amounts of dust accumulated on these bodies also reinforces the idea that the solar system has not existed for the long periods of time that are commonly accepted.

Common counterarguments:

Some argue that certain processes, such as lunar volcanic activity or meteorite impacts, could redistribute or remove dust, explaining why there isn't a lot of accumulated dust.

However, on the Moon, these processes are limited and do not fully explain the absence of a significant layer of interplanetary dust.

Bottom line: The amount of interplanetary dust accumulated on the Moon and other solar system bodies is incompatible with the idea that the solar system is billions of years old.

Sources supporting the argument:

Simpson, J. A., & Bowhill, S. A. (1975). Interplanetary Dust Particles Collected on the Lunar Surface. Science, 188(4184), 1295-1296.

Brownlee, D. E., & Hemenway, C. L. (1975). Lunar Dust: A Cosmic Perspective. Proceedings of the Lunar and Planetary Science Conference, 6, 3881-3888.

Geiss, J. (1973). The lunar atmosphere and the dust flux. Philosophical Transactions of the Royal Society of London. Series A, Mathematical and Physical Sciences, 274(1239), 271-280.

12) Alignment of the planets:

The alignment and motions of the planets in the solar system provide evidence that questions the times proposed by the evolution of the solar system.

According to the evolutionary model, the planets of the solar system formed about 4.6 billion years ago from a disk of dust and gas orbiting the Sun. Something that contradicts many physical laws, since it is assumed that they were compacted due to certain forces, but it turns out that these depend on the mass and the square of their distances, and well that is another physical topic that is not the idea of this book. For this reasoning and a similar one by Father Carreira, he even called Hawking stupid, a confrontation between two theoretical physicists, one a believer and the other a believer.

During that long period, gravitational interactions between the planets should have caused certain misalignments and chaotic changes in their orbits.

Evidence observed:

Despite the long-proposed time for the existence of the solar system, the orbits of the planets remain relatively stable and organized.

The chaotic misalignments that one would expect due to prolonged gravitational interactions between planetary bodies are not observed.

This fact suggests that the gravitational forces between the planets, over billions of years, have not caused the effects we would expect.

In particular, the planets Uranus and Neptune exhibit orbital inclinations that do not fit well into the standard evolutionary model. Uranus, for example, has an extreme axial tilt of nearly 98 degrees, suggesting that its orbital history doesn't follow a pattern that would have been expected if it had been orbiting for billions of years in such an old and stable solar system.

Orbital stability:

Studies on the orbital dynamics of the planets show that long-term gravitational interactions should have caused further disruption in their orbits.

If the solar system really were billions of years old, we should see significant effects, such as chaotic resonances or collisions between smaller bodies, in addition to disordered orbits.

What we observe, however, is relative stability in the orbits of the planets, suggesting that the solar system is younger than proposed.

The orbits remain organized and regular, which is incompatible with a system that has existed for so long under the effect of continuous and cumulative gravitational influences.

Questioned models:

In addition to orbits, the orbital inclinations of some planets, such as those of Uranus and Neptune, remain an enigma to evolutionary models of planet formation.

These unusual inclinations are not easily explained within the framework of the solar system formation model in 4.6 billion years. The evolutionary model suggests that collisions and chaotic events must have affected these planets in their early stages of formation. However, the details about how these events may have occurred, and their effects are unclear in the standard model, and would seem to be accepted by faith.

Implications for the age of the solar system:

The stability observed in the planets' orbits, coupled with anomalous orbital inclinations, suggests that the solar system has not existed for billions of years. The cumulative effects of gravity and collisions that we would expect to observe in such an ancient system do not manifest in the expected way. This opens up the possibility that the solar system has a much more recent history than current models suggest.

Suggested sources to strengthen the argument:

Chambers, J. E. (1999). A hybrid simplistic integrator that permits close encounters between massive bodies. Monthly Notices of the Royal Astronomical Society, 304(4), 793-799.

Laskar, J. (1989). A numerical experiment on the chaotic behavior of the solar system. Nature, 338(6212), 237-238.

Gomes, R., et al. (2005). Origin of the cataclysmic Late Heavy Bombardment period of the terrestrial planets. Nature, 435(7041), 466-469.

Part II: Geological Evidence

13) The erosion of the continents:

The erosion of continents is a natural process that occurs continuously due to the action of water, wind, and other environmental factors. Every year, it is estimated that about 25 billion tons of sediment are washed away from the continents and deposited in the oceans.

At current rates of erosion, continents should have been weathered to sea level in less than 20 million years, even considering the tectonic processes that lift mountains and other high ground.

However, there are still high mountains and large expanses of dry land, which raises serious doubts about the chronology proposed by evolutionary models.

Evidence observed:

Erosion rates: It has been estimated that about 25 billion tons of sediment are washed annually from the continents into the oceans. At this rate, the mainland should be completely eroded in about 20 million years. If the Earth were billions of years old, as evolutionary theory suggests, the continents would have already been eroded to sea level several times.

Sediments on the ocean floor: Given the enormous volume of sediment deposited in the oceans each year, we would expect to find layers of sediment several kilometers deep on the ocean floor if the Earth were billions of years old.

However, what is observed is a relatively thin layer of sediment, only a few hundred meters in most cases.

This suggests that the time during which continents have been eroding and depositing sediment in the oceans is much shorter than predicted in conventional geological models.

Mountains still existing: Despite erosion rates and the action of the elements over time, tall mountains and elevated landscapes still exist, which is inconsistent with an erosion process that should have completely worn them away if the Earth were billions of years old.

Although tectonic processes raise mountains, the balance between erosion and uplift is not enough to sustain mountains for such long periods of time.

Implications for the age of the Earth:

If erosion had been going on for billions of years, we should observe complete weathering of the continents and a large accumulation of sediment on the ocean floor.

The fact that this is not what we observe suggests that the Earth has not existed in its current state for billions of years. Rather, the balance between erosion and sedimentation indicates that the continents are much younger than is commonly assumed.

Common counterarguments:

Some geologists propose that tectonic processes that raise mountains counteract erosion, but this explanation has its limitations.

Although tectonic forces lift mountains, these same mountains should have been eroded several times if the Earth was really billions of years old.

In addition, the relatively thin layer of sediment at the bottom of the oceans is difficult to explain if the Earth has been accumulating sediment for so long.

The current rates of erosion of the continents and the scarcity of sediment at the bottom of the oceans are not consistent with an Earth billions of years old. Instead, they suggest a much younger world, in which erosion has not had enough time to completely wear away the continents or to accumulate the levels of sediment that we would expect to find if the Earth were as old as evolutionary theory postulates.

Suggested sources supporting the argument:

Milliman, J. D., & Syvitski, J. P. M. (1992). Geomorphic/tectonic control of sediment discharge to the ocean: the importance of small mountainous rivers. Journal of Geology, 100(5), 525-544.

Summerfield, M. A., & Hulton, N. J. (1994). Natural controls of fluvial denudation rates in major world drainage basins. Journal of Geophysical Research, 99(B7), 13871-13883.

Hay, W. W. (1998). Detrital sediment fluxes from continents to oceans. Chemical Geology, 145(3-4), 287-323.

14) Corrosion in Niagara Falls:

Niagara Falls is a natural phenomenon of constant erosion. The waterfall, which lies between Canada and the United States, is eroding and receding into Lake Erie at an average rate of about one meter per year.

By calculating the total distance, the falls have receded, and taking into account their current rate of erosion, the falls are estimated to be less than 10,000 years old.

This estimate is significantly less than the millions of years that evolutionary chronology suggests, but it is more in line with a biblical chronology.

Evidence observed:

Erosion rate: The erosion of Niagara Falls has been observed and measured for years, and the average retreat is estimated to be about one meter per year. This erosion is mainly due to the force of the water, which hits the bedrock and gradually wears it away, causing the falls to recede into Lake Erie.

Distance traveled: Since their origin, the falls have receded approximately 11 kilometers from the place where they originally began to erode.

If we use the current erosion rate of one meter per year, we can calculate that the falls have been eroding for about 9,000 years.

Changes in erosion rate: Although some studies suggest that the rate of erosion has varied over time, even if we take these fluctuations into account, the estimated maximum time for the formation and retreat of the falls is still much less than the millions of years proposed by geological evolutionary models.

Implications for the age of the Earth:

If Niagara Falls were really only 9,000 years old, this would be a strong indication that some of the geological processes affecting Earth have not been occurring for millions of years as the evolutionary model suggests.

This calculation aligns with a biblical chronology, which suggests a younger world.

In addition, the young age of Niagara Falls means that the geological conditions that have allowed its formation are relatively recent in Earth's history. This could be related to recent climate or tectonic changes, rather than ancient processes that have been going on for millions of years.

Common counterarguments:

Some geologists suggest that erosion rates may have been slower in the past or that different factors, such as glacier formation and subsequent melting, have affected the rate of retreat of the falls.

However, even considering these factors, the estimated time for the retreat of the falls is still much shorter than the millions of years proposed in evolutionary models.

Suggested sources for research:

Gilbert, G. K. (1906). Niagara Falls and Their History. Bulletin of the Geological Society of America, 16(1), 157-186.

Tinkler, K. J., & Parish, C. P. (1998). The Great Cataract at Niagara: A Geological Perspective on Niagara Falls. Geography Review, 12(3), 25-34.

Greenberg, B. (2006). Erosion and Evolution of the Niagara Escarpment. Geological Journal, 41(5), 543-558.

15) The Mississippi River Delta:

River deltas, such as the Mississippi Delta, are formed by the accumulation of sediment carried by rivers and deposited at their mouths. This process occurs consistently over time.

If deltas had been forming for millions of years, as evolutionary theory suggests, their size and amount of sediment would be much larger than what is currently observed.

However, current measurements of the growth of the Mississippi Delta and other river systems suggest a much younger age for these geologic systems.

Evidence observed in the Mississippi Delta:

Sediment accumulation rate: The Mississippi River Delta is estimated to accumulate sediment at a rate of approximately 300 million tons per year. Over time, this accumulation forms the large expanses of land at the mouth of the river, known as the delta.

Given the current volume of sediment in the Mississippi Delta, calculations suggest that this delta is less than 30,000 years old.

This is inconsistent with the idea that the delta has been forming for millions of years, as the volume of sediment would be much larger if that were the case.

Comparison with other deltas:

Nile Delta: Although studies of the Nile Delta are not as detailed as those of the Mississippi, research indicates that this delta has been growing for approximately 5,000 to 7,000 years, coinciding with the beginning of human sedentary lifestyles and the emergence of the first civilizations in Egypt. The rate of sediment accumulation in the Nile is also incompatible with a formation of millions of years.

Euphrates River: Unlike the Mississippi or the Nile, the Euphrates does not have a delta of great magnitude, as it flows into the Persian Gulf next to the Tigris.

However, sedimentation studies in the Mesopotamian region indicate that sediment accumulation patterns are not consistent with millions of years old.

Instead, these studies point to more recent formation, likely within the last 10,000 years, which coincides with the historical record of early Mesopotamian civilizations.

Amazon Delta: Data on sediment accumulation in the Amazon, the world's largest river system, is also inconsistent with a formation period of millions of years.

Studies suggest that the volume of sediment in its delta is much smaller than what would be expected if it had been forming for so long, reinforcing the idea that these systems are more recent than evolutionary models propose.

Implications for the age of the Earth:

If river deltas like the Mississippi, Nile, and Amazon were really millions of years old, we should see a much larger accumulation of sediments and deltas of disproportionately large size. However, the current amount of sediment in these deltas suggests that their formation has occurred in a much shorter period, pointing to a much younger Earth than conventional evolutionary models suggest.

Furthermore, this data matches evidence that the first human civilizations began settling near these rivers less than 10,000 years ago, reinforcing the chronology of a more recent Earth.

Common counterarguments:

Some geologists argue that sedimentation rates may have fluctuated over time due to weather events, changes in sea level, or variations in the amount of eroded material. However, even if these fluctuations are taken into account, the observed amounts of sediment in river deltas are inconsistent with the millions of years proposed by evolutionary models.

Suggested sources to support the argument:

Roberts, H. H. (1997). Dynamic changes of the Holocene Mississippi River delta plain: The delta cycle. Journal of Coastal Research, 13(3), 605-627.

Frazier, D. E. (1967). Recent deltaic deposits of the Mississippi River: their development and chronology. Transactions of the Gulf Coast Association of Geological Societies, 17, 287-315.

Stanley, D. J., & Warne, A. G. (1993). Nile delta: Recent geological evolution and human impact. Science, 260(5108), 628-634.

16) Formation of stalactites and stalagmites:

Stalactites (which hang from the ceiling of caves) and stalagmites (which rise from the ground) are mineral formations that develop through the slow deposition of minerals dissolved in dripping water.

Traditionally, it has been taught that these structures require thousands or millions of years to reach their current size. However, recent observations have shown that these formations can grow at a much faster rate under certain conditions, calling into question the need for extensive geological times for their formation.

Evidence observed:

Growth in modern structures: Stalactites and other mineral formations have been observed forming in man-made structures, such as bridges, tunnels, and buildings, over a span of decades. These structures, which were built in the last century, show how the right conditions of dripping and deposition of minerals can lead to the formation of remarkable stalactites in relatively short times.

For example, in railway tunnels and under bridges built in the mid-20th century, stalactites several centimeters to several meters long have been observed, indicating that mineral growth may be much faster than traditionally thought.

Accelerated growth conditions: Under favorable conditions, where the flow of water and the concentration of dissolved minerals is high, stalactites and stalagmites can grow faster than normal. These conditions include a constant rate of dripping water and an atmosphere that favors the deposition of minerals such as calcite ($CaCO_3$).

Some studies have documented stalactite growth rates of between 0.1 and 2.5 millimeters per year, meaning that large formations could have developed in less than 4,400 years, *a time compatible with biblical chronology based on the Flood.*

Observations in natural caves: Although stalactite and stalagmite formations in natural caves are often considered very old, the growth rates observed in these caves can vary considerably depending on local conditions.

In some caves, the growth observed in recent times indicates that the current formations could have developed in a much shorter period than traditional models of thousands of years suggest.

Impact of the Flood: According to creationist chronology, the biblical Flood, which would have occurred approximately 4,400 years ago, may have created conditions that favored the rapid deposition of minerals.

The increase in geothermal activity (especially in Peleg's time), the heating and cooling of groundwater, and the change in water levels in caves could have accelerated the formation of stalactites and stalagmites.

The Bible mentions Peleg in the book of Genesis 10:25, where it says: "To Heber were born two sons: the name of the one was Peleg, because in his days the land was divided..." The name Peleg is associated with the Hebrew word meaning "to divide," which has led many to interpret that in Peleg's day some splitting event occurred, such as the division of the nations at the tower of Babel or even a geological division of the Earth. Peleg's time would be approximately **4271 years from** today.

The Bible is not meant to be a scientific or geological treatise, but its primary purpose is to reveal the nature of God, His relationship to humanity, and His plan of salvation through redemptive history. Therefore, when he mentions historical or natural events, such as the division of the Earth in Peleg's day (Genesis 10:25), he does so in a way that aligns with the larger purpose of conveying God's spiritual message.

It is important to remember that the Bible is not intended to provide a detailed chronology of geological or scientific events. Their focus is on revealing how God interacts with His creation and how He carries out His redemptive plan. Details about natural or catastrophic events, such as the possible splitting of continents, are not the primary focus of Scripture. Instead, the Bible focuses on the revelation of God's character and His work in the salvation of mankind through His people and his redeemed.

So, while there may be hints of catastrophic events such as the separation of the lands in the age of Peleg, the Bible does not abound in detail on these issues because its purpose is much deeper: it is a treatise of salvation and a manifestation of how God acts in human history to redeem mankind and carry out his sovereign plan.

But the flood or the possible division of the earth into continents as it seems that a Pangaea initially existed without a catastrophic event would have generated major changes in water systems, allowing the flow of mineral-rich water to significantly accelerate deposition in caves, leading to the rapid formation of the large structures we see today.

Implications for the age of the Earth:

The observed growth rates of modern stalactites and stalagmites call into question the need for thousands or millions of years for large structures to form. Instead, evidence suggests that, under the right conditions, these formations can develop in a relatively short period of time, consistent with an Earth much younger than conventional evolutionary models propose.

Common counterarguments:

Some geologists argue that stalactites in natural caves grow more slowly than in modern structures due to differences in mineral composition and climatic conditions.

However, even with these variations, the growth rates observed today are much faster than originally believed, suggesting that it is not necessary to postulate extensive geological times to explain these formations.

Stalactites and stalagmites can form much faster than has been traditionally taught. Recent studies of mineral formations in modern structures and in natural caves suggest that these formations may have grown in less than 4,400 years, supporting a more recent chronology of the Earth, in line with the biblical interpretation of geological history. This challenges the notion that these structures require millions of years to develop and instead reinforces the possibility of a younger Earth.

Hill, C. A., & Forti, P. (1997). Cave Minerals of the World. National Speleological Society.

Shopov, Y., et al. (1994). Luminescence of cave minerals. Cave Minerals of the World, 2nd edition.

Curl, R. L. (1966). Cave morphology and the rate of cave forming processes. National Speleological Society Bulletin, 28, 1-14.

17) The oldest coral reef:

The growth of coral reefs, such as the Great Barrier Reef in Australia, which is the largest reef in the world, provides important information about the age of these ecosystems and, potentially, the age of the Earth itself.

Coral reefs grow at a measurable rate that varies depending on water conditions, temperature, sunlight, and other factors.

Based on the observed growth rate, it is estimated that the world's largest coral reefs could have formed over a period of about 4,200 years.

Evidence observed:

Coral growth rate: Coral reefs, such as the Great Barrier Reef, grow at a rate of between 1 to 3 cm per year on average, depending on environmental conditions, such as water temperature and sunlight availability.

At this rate of growth, studies estimate that even the world's largest reefs would have taken less than 4,200 years to reach their current size.

Oldest coral reefs: Currently, no evidence has been found of coral reefs that are more than 4,200 years old, raising questions about the age of the Earth from a creationist perspective.

If the Earth were millions of years old, as evolutionary theory suggests, we should observe older reefs or evidence of multiple cycles of reef growth, destruction, and regeneration over those millions of years.

However, there are no living reefs that directly proved to have existed for millions of years. The largest and most complex reefs observed today were formed in a much shorter period of time, which appears to be inconsistent with the long geological times proposed by the theory of evolution.

Growing conditions: Coral reefs are very sensitive to environmental conditions, and their growth can be affected by events such as changes in sea level, pollution, and temperature fluctuations. Despite these variations, the fact that the world's largest reefs can form in less than 4,200 years suggests that millions of years don't need to be postulated to explain their existence.

In addition, the lack of evidence of previous coral reef cycles dating back millions of years is an indication that these ecosystems may have begun to form after a catastrophic event, such as the biblical Flood, which, according to biblical chronology, occurred approximately 4,400 years ago.

Implications for the age of the Earth:

The estimate that the world's oldest coral reefs formed over a period of 4,000 to 4,200 years is significant from a creationist perspective. If Earth were millions of years old, we would expect to find much larger reefs or evidence of older reefs that have gone through multiple cycles of growth and destruction.

The lack of such evidence and the relatively young age of today's reefs is more in line with biblical chronology, which suggests a much younger Earth.

In addition, the rapid growth of stalactites, stalagmites, and other natural geological processes also accords with this idea, defying the long time periods postulated by evolutionary theory.

Common counterarguments:

Some scientists argue that coral reefs have been destroyed and rebuilt several times due to weather events or changes in sea level, which could explain the lack of older reefs.

However, there is no conclusive evidence from multiple previous coral reef cycles that can support a million-year model.

Today's reefs show steady growth over the past few thousand years, raising questions about the long-term geological scenario.

Hopley, D. (1982). The geomorphology of the Great Barrier Reef: Quaternary development of coral reefs. John Wiley & Sons.

Adey, W. H., & Macintyre, I. G. (1973). Crustose coralline algae: A re-evaluation in the geological sciences. Geological Society of America Bulletin, 84(3), 883-904.

Veron, J. E. N. (2008). A Reef in Time: The Great Barrier Reef from Beginning to End. Harvard University Press.

18) Marine sediments:

Argument: If Earth were billions of years old, we should find layers of sediment at the bottom of the oceans that reached several kilometers deep.

However, what we observe today are layers of sediment that only reach a few hundred meters.

According to geological studies, ocean sediments accumulate at an average rate of **20 millimeters per thousand years**, although this rate can vary depending on the location and environmental conditions. However, even at a conservative rate, millions of years of accumulation should have resulted in **much thicker deposits** than is observed today.

- For **1 million years**:

    o   At a rate of 20 mm per 1,000 years, approximately **20 meters** of sediment would accumulate.

- For **4.6 billion years** (which is the proposed age for Earth):

    o   At the same rate, approximately **92,000 meters**, or **92 kilometers** of sediment, would have accumulated.

This calculation shows that, if the Earth were billions of years old, we should find extremely thick layers of ocean sediments, which is not observed in reality, where we only find sediments of a few hundred meters.

The current average sediment on the ocean floor is generally a few hundred meters (e.g., between 300 and 400 meters of sediment in many areas), although this amount can vary depending on the specific location in the oceans and local geological conditions.

If we consider this rate of accumulation and the average thickness of sediments of **300 to 400 meters**, we can estimate the time it would have taken for that amount of sediment to accumulate:

1. **Calculation of time needed**:

    o   **300 meters of sediment** at a rate of 20 mm every 1,000 years implies:

        ▪   **300,000 mm of sediment / 20 mm per 1,000 years = 15 million years**.

    o   **400 meters of sediment** at the same rate would imply:

        ▪   **400,000 mm of sediment / 20 mm per 1,000 years = 20 million years**.

Implications of the biblical Flood:

From a creationist perspective, it is postulated that the **biblical Flood** may have been a catastrophic event that caused a massive and rapid accumulation of sediment around the world. During that event, accumulation rates would have been much higher than those observed in normal times, which would explain the large amount of sediment accumulated in a short period of time.

This model suggests that, instead of millions of years of slow accumulation, most of today's ocean sediments could have been deposited during and shortly after the Flood, dramatically accelerating accumulation in the deep ocean.

At the current accumulation rate of 20 mm per thousand years, the observed sediment of 300 to 400 meters would have taken between 15 and 20 million years to form.

However, if we consider a catastrophic event such as the Flood, most of the sediment could have accumulated rapidly, in a much shorter period, which would support a more recent chronology of the Earth, consistent with a biblical interpretation of the Flood as primarily responsible for this sediment accumulation.

This model challenges the idea of a slow, progressive accumulation of sediment over billions of years and raises the possibility that catastrophic events may have played a key role in the formation of the sediment layers we see in the oceans today.

This discrepancy suggests that the sediment accumulation time could be **much shorter** than conventional geological models indicate.

This is inconsistent with the idea of an Earth that has existed for billions of years, as it would have accumulated much more sediment during that time.

Common counterarguments:

Some geologists suggest that geological processes, such as the subduction of tectonic plates, could be recycling sediment accumulated on the ocean floor, which would explain the lack of deep layers.

However, even with this recycling, the observed amount of sediment is still much lower than would be expected if the Earth were billions of years old.

Milliman, J. D., & Meade, R. H. (1983). World-wide delivery of river sediment to the oceans. The Journal of Geology, 91(1), 1-21.

Hay, W. W. (1994). Pleistocene-Holocene fluxes are not the Earth's norm. Geological Society, London, Special Publications, 95(1), 261-283.

Emery, K. O., & Uchupi, E. (1972). Western North Atlantic Ocean: sedimentary evolution and Cenozoic history. Geological Society of America Bulletin, 83(1), 71-88.

19) Pressure in oil tanks:

One of the phenomena observed in many oil wells is the extremely high pressure within underground reservoirs.

According to conventional theories, these deposits were formed millions of years ago. However, the pressure found in many of these deposits is not compatible with that long time, due to the permeability of the surrounding rocks.

Evidence observed:

High pressure in reservoirs: Many oil wells show very high levels of internal pressure when drilled, suggesting that the fluids and gases within the reservoir have been under confinement for a relatively short period.

If the oil deposits really were millions of years old, as proposed in evolutionary theories, the internal pressure should have decreased considerably due to the natural permeability of the rocks surrounding the deposits.

This is because, over time, fluids and gases tend to slowly escape through rocks.

Rock permeability: The rocks surrounding oil deposits have limited permeability, which means they allow fluid flow, albeit slowly.

Over millions of years, this permeability would allow gas and oil to seep out and internal pressure to dissipate significantly.

However, in reality, it is observed that the deposits still retain high pressure, indicating that these fluids have been trapped for a much shorter period of time, rather than millions of years.

Pressure dissipation calculations: Studies on rock permeability indicate that pressure in a closed reservoir should dissipate in much less time than conventional theories suggest.

The properties of the rocks suggest that they could not have retained fluids at high pressure for millions of years without there being a leakage or considerable reduction in pressure.

This reinforces the idea that oil has not been trapped in these deposits for millions of years, but that the formation and confinement of oil occurred on a much shorter timescale.

Implications for the age of the Earth:

The presence of high pressure in many oil deposits suggests that these natural resources formed and were trapped recently, rather than millions of years ago.

If the Earth were as old as evolutionary models postulate (billions of years), oil deposits would have lost their pressure long ago due to the permeability of the rocks.

This phenomenon is most consistent with a young Earth, in which oil and gas were formed and trapped in a relatively short period of time, possibly as a result of a catastrophic event, such as the biblical Flood.

In this model, sediments and geological conditions would have confined the oil in recent rock formations, maintaining pressure until today.

Common counterarguments:

Some geologists suggest that the high pressure in oil reservoirs can be explained by the presence of impermeable layers that prevent the escape of fluids.

However, even with these impermeable layers, the internal pressure should have dissipated considerably within millions of years.

The rocks surrounding the deposits are not completely impermeable, meaning that the pressure should eventually have been reduced, which is not what is observed.

Bradley, W. H. (1973). Pressure in petroleum reservoirs and its dissipation. Journal of Petroleum Technology, 25(01), 23-27.

Magara, K. (1978). Compaction and Fluid Migration: Practical Petroleum Geology. Elsevier.

Nelson, P. H. (1994). Permeability-porosity relationships in sedimentary rocks. Log Analyst, 35(03), 38-62.

20) Evidence in Tectonic Plates:

The movement of tectonic plates is generally interpreted as a very slow process that has occurred over millions of years, but some creationist geologists propose an alternative model that suggests that these movements could have been much faster and catastrophic, especially in the context of events such as Noah's Flood.

This model argues that geological processes, such as the formation of mountains, mountain ranges, and geological faults, could have occurred in a much shorter period of time than conventional models suggest.

Evidence observed:

Marine fossils on mountain tops:

Marine fossils have been found on the tops of many mountains, such as the Himalayas. These findings indicate that these mountain formations were once submerged underwater. The question that arises is how these huge mountain ranges could have risen from the ocean floor to today's heights in such a short time.

Creationist models suggest that the biblical Flood and associated catastrophic events could have caused a rapid uplift of mountains and mountain ranges, pushing marine sediments to great heights in a short period.

Geological faults and mountain ranges: Huge geological faults, such as the San Andreas Fault in California, are interpreted in conventional models as the result of millions of years of slow movement of tectonic plates.

However, creationists propose that these faults could have formed quickly during a global catastrophic event.

The rapid shift of tectonic plates in a young Earth model could explain the formation of vast mountain ranges, such as the Andes and the Himalayas, in a short period of time rather than the millions of years proposed in evolutionary models.

Tectonic movement rates: The movement rates of tectonic plates are currently measured in centimeters per year, which is interpreted as an extremely slow process.

However, some creationist geologists suggest that, during a catastrophic event such as the Flood or the splitting of Pangaea in Peleg's time, these rates could have been much faster, causing the rapid formation of today's geological features.

Huge faults and compression of tectonic plates could have occurred at a much higher rate in the past due to catastrophic forces, accelerating processes that we observe today as slow.

Models of Catastrophizing Tectonics: Creationists propose a model of Catastrophic Plate Tectonics, in which the rapid movement of plates during the Flood explains many geological phenomena that are observed today, such as the rapid elevation of mountains and the formation of large mountain ranges.

This model is consistent with the young Earth view, as it suggests that these events occurred in a much shorter period, rather than the slow gradual changes assumed by the evolutionary view.

Implications for the age of the Earth:

The evidence of marine fossils on mountaintops, along with the huge geological faults and mountain ranges, can be interpreted as the result of a massive catastrophic event, such as the biblical Flood. By the way, demonstrating The Flood, or the evidence in its favor is another matter. That would be another book. But this one borders on the evidence in favor of the Flood.

While Plate Tectonics is generally seen as a slow process that occurs over millions of years, this model suggests that these movements may have been rapid and cataclysmic in the recent past.

This model is consistent with a young Earth, as it proposes that many of the geological processes we observe today occurred rapidly in the context of a global catastrophic event. This challenges the conventional view of long periods of time and suggests that the Earth is not millions of years old.

Common counterarguments:

Evolutionary geologists argue that tectonic forces have acted steadily for millions of years, and that marine fossils in mountains are simply the result of the slow ascent of geological formations due to the gradual movement of tectonic plates.

However, the Catastrophic Plate Tectonics model suggests that these processes may have occurred much more rapidly in the past, in a scenario of accelerated geological change.

Baumgardner, J. R. (1994). Runaway subduction as the driving mechanism for the Genesis Flood. Proceedings of the Third International Conference on Creationism.

Snelling, A. A. (2009). Earth's Catastrophic Past: Geology, Creation, and the Flood. Institute for Creation Research.

Austin, S. A., & Wise, K. P. (1994). The pre-Flood/Flood boundary: As defined in Grand Canyon, Arizona and eastern Mojave Desert, California. Proceedings of the Third International Conference on Creationism.

21) Erosion of volcanoes:

The volcanic activity observed today suggests that Earth could be much younger than postulated in conventional geological models, which propose that Earth is about 4.6 billion years old.

If the Earth were as old as claimed, the weathering of volcanoes and the dissipation of internal heat should have already caused a significant decrease or complete disappearance of volcanic activity.

Evidence observed:

Continuous volcanic activity: Around the world, active volcanoes are observed that continue to release large amounts of magma, gases, and heat from the Earth's interior.

This activity has been constant throughout recorded history, and significant volcanic eruptions can still be observed today.

Conventional models hold that this activity is a remnant of the Earth's early formation.

However, after 4.6 billion years, we would expect that Earth's internal heat would have dissipated enough that volcanic activity would have greatly decreased or even disappeared.

Volcano weathering: Volcanoes are susceptible to erosion and wear and tear from environmental factors such as wind, water, and tectonic activity.

If the Earth were as old as proposed, many of the volcanoes that are active today should have been completely eroded.

For example, some of the world's most iconic volcanoes, such as Mount St. Helens or Mount Fuji, still show great activity and a fairly defined shape.

If these volcanoes were millions of years old, they should have been eroded into simple mounds or rocky debris, rather than active, well-formed structures.

Earth's internal heat: The internal heat that fuels volcanic activity comes in part from the radioactive decay of elements in Earth's core and mantle.

Although these processes can release heat over long periods, Earth has been radiating heat into space for billions of years. If Earth were that age, much of this heat would have already dissipated, significantly reducing volcanic activity.

The presence of active volcanoes today suggests that the Earth's interior still retains large amounts of thermal energy, which is more consistent with a younger Earth than with one that is billions of years old.

Eruption rates: Some volcanoes, such as Mount Kilauea in Hawaii, have been observed to have been erupting continuously for decades, releasing massive amounts of volcanic material.

Over millions of years, this number of eruptions should have significantly reduced volcanic activity, which is not what is observed today.

Implications for the age of the Earth:

If Earth were really 4.6 billion years old, we would expect to observe a significant decrease in volcanic activity due to internal heat dissipation and erosion of volcanoes.

However, the fact that many volcanoes are still active and maintain their structural shape, despite the expected millions of years of erosion, suggests that Earth could be much younger.

This phenomenon is consistent with a young Earth, where the geological processes that generate heat and volcanic activity have occurred on a much shorter timescale. Instead of heat dissipation that has lasted for billions of years, Earth still holds enough internal thermal energy to power active volcanoes today.

Common counterarguments:

Some geologists argue that volcanic activity is sustained by the continuous recycling of material in the Earth's mantle, which allows the Earth to maintain its internal heat for longer than would be expected.

However, even with this recycling, the amount of heat that would have dissipated over billions of years should have reduced volcanic activity on a significant scale.

The current erosion processes should also have worn away many of the volcanoes that are still active today.

Fisher, R. V., & Schmincke, H.-U. (1984). Pyroclastic Rocks. Springer-Verlag.

Wood, C. A., & Kienle, J. (1990). Volcanoes of North America: United States and Canada. Cambridge University Press.

Hamblin, W. K., & Christiansen, E. H. (1995). Exploring the Planets. Prentice Hall.

22) Salinity of the oceans:

The concentration of salt (sodium) in the oceans provides a strong argument against the idea of a billion-year-old Earth. If the oceans had existed during that time, they would have been much saltier than they are today, which would have made life impossible in them.

According to scientific studies conducted by Steve Austin and Russell Humphreys in 1991, the amount of sodium in the oceans suggests a maximum age of 62 million years under the most favorable conditions for slow accumulation, although many studies suggest an even younger age for the oceans.

Evidence observed:

Sodium ingresses into the oceans: Sodium (salt) enters the oceans through processes such as rock erosion, rivers that carry dissolved minerals, and other natural sources.

This income is continuous and has been occurring for as long as the oceans have existed.

If the oceans have been around for billions of years, the concentration of salt should have reached much higher levels than are observed today, making life in today's oceans extremely difficult, or even impossible.

Limited sodium output: Although some sodium leaves the ocean through processes such as mineral deposit formation and evaporation, the rate of output is much slower than the rate of input.

This means that the concentration of salt in the oceans should be increasing steadily over time.

Austin and Humphreys calculated the rate of sodium inflow and outflow, and even taking into account the most conservative conditions (minimum input and maximum output of sodium), they determined that the oceans cannot be more than 62 million years old.

This figure is much lower than the billions of years proposed by evolutionary models.

Current salt concentration: The current salinity of the oceans is about 3.5% on average.

If the oceans had existed for billions of years, this concentration should be significantly higher, probably high enough that life as we know it could not have survived.

The fact that salinity levels in the oceans still allow for the existence of marine life is an indicator that the oceans are not billions of years old.

While evolutionists argue that life began in salty oceans, the exact amount of salinity at that time is still up for debate, adding further uncertainty to their theories.

Austin and Humphreys Studies: In their study titled *"The Sea's Missing Salt,"* Austin and Humphreys evaluated both sodium inflow and sodium outflow into the oceans. They considered the possibility that the oceans had started out without any salt and calculated how long it would have taken to reach today's salinity. Even under the most favorable conditions for evolutionary models, the salinity of the oceans indicates that the Earth cannot be more than 62 million years old.

The research also showed that the concentration of sodium accumulated in the oceans does not match the chronology of fossils that evolutionists use to date the Earth and oceans. This raises important questions about the validity of conventional dating methods.

Implications for the age of the Earth:

If the oceans were billions of years old, their salt concentration would be much higher than it is today. The salt levels observed today suggest that the oceans are much younger, and that they have not existed for as long as evolutionary models propose.

Austin and Humphreys' research indicates that, under the conditions most favorable to evolutionary models, oceans cannot be more than 62 million years old.

What's more, life as we know it couldn't have survived in oceans with higher salinity levels for extended periods of time.

Common counterarguments:

Some scientists suggest that the rate of sodium output might have been higher in the past, which would explain the current concentration of salt in the oceans.

However, observed evidence indicates that sodium outflow is significantly slower than its inflow, suggesting that salinity should have increased much more if the oceans were indeed billions of years old.

In addition, evolutionists disagree about the conditions under which life could have arisen in the oceans, which introduces more uncertainty into their theories.

Austin, S. A., & Humphreys, D. R. (1991). The Sea's Missing Salt: A Dilemma for Evolutionists. Creation Research Society Quarterly, 17-33.

Hay, W. W., et al. (2006). The Mass of Salt in the Oceans: A Challenge for Evolutionary Models. Geological Society of America Bulletin.

Milliman, J. D., & Meade, R. H. (1983). World-wide delivery of river sediment to the oceans. The Journal of Geology, 91(1), 1-21.

23) The Rocky Mountains:

The Rocky Mountains are one of the most iconic mountain ranges in North America.

Evolutionary geological models hold that these mountains are hundreds of millions of years old.

However, the degree of natural erosion observed in the Rockies is inconsistent with such an advanced age. If they really were that old, they should have suffered much greater wear and tear due to erosion processes, which raises doubts about their true origin and age.

Evidence observed:

Continuous erosion in mountains: Erosion is a constant phenomenon caused by wind, rain, ice, and other climatic factors that wear down land surfaces over time. This process should have significantly affected the Rocky Mountains if they were actually hundreds of millions of years old.

It is estimated that mountain erosion reduces their height steadily, depending on weather conditions, wind patterns, and precipitation. Over millions of years, erosion should have greatly reduced the height and prominence of the Rockies.

Conservation of geological features: What we currently observe is that the Rocky Mountains still retain prominent and relatively well-defined features, which is not consistent with the level of erosion that would be expected after millions of years of natural weathering.

Despite the forces of erosion that have acted over long periods of time, the Rockies and other mountain ranges show a level of conservation that is inconsistent with a chronology of hundreds of millions of years.

If the mountains were that age, they should have been weathered significantly more, leaving only eroded and low formations.

Incompatibility with evolutionary models: Conventional geological models suggest that the Rocky Mountains are around 55 to 80 million years old, based on the theory of plate tectonics and mountain formation.

However, the amount of erosion observed does not support such a long timescale.

We can rely on studies of the average rate of mountain erosion, although this rate varies depending on climate, geology, and other factors. However, in general, mountain erosion rates today have been estimated to be in the range of 0.01 mm to 1 mm per year, depending on the region and specific conditions.

Rocky Mountain Erosion Rate:

If we consider an **average erosion rate** of **0.1 mm per year**, which is an intermediate value within the mentioned range, we can calculate the **accumulated wear** in millions of years.

1. **Erosion rate:**

    o At **0.1 mm per year**, in **a million years**, a mountain would lose approximately **100 meters** in height due to erosion.

2. **Rocky Mountain Height:**

    o If the Rocky Mountains were between **55 and 80 million years old** (according to conventional geological models), they would have lost between **5,500 and 8,000 meters** in height due to erosion, according to this rate.

3. **Rocky Mountain Current Height:**

    o The average height of the **Rocky Mountains** varies between **2,000 and 4,400 meters**. However, if the Rockies had lost between **5,500 and 8,000 meters** due to erosion in the last **55 to 80 million years**, this would indicate that the mountains should have been significantly higher in their original formation, possibly reaching heights of between **7,500 and 12,000 meters**.

4. **Incompatibility with erosion**:

    o Given the level of erosion expected at this time, the **Rocky Mountains** should have been **completely eroded** or reduced to small hills. The fact that they still maintain such a significant height and prominence suggests that their **formation is much more recent**.

    o Even if the rate of erosion had been lower in the past, we would still expect a much higher level of wear than is seen today.

Implications:

If the Rocky Mountains were really between **55 and 80 million years old**, as conventional geological models suggest, the erosion process should have reduced them significantly. If the mountains had had a much higher initial height (around **10,000 to 12,000 meters**), they would have been the highest mountains in the world at the time. However, the **amount of erosion observed** today does not agree with such a long chronology.

This reinforces the idea that the Rocky Mountains could have formed **much more recently**, and that the erosion rates observed today are **incompatible** with a timescale of **tens of millions of years**.

Based on **current erosion rates**, the **Rocky Mountains** should have been **completely eroded** if they were as old as conventional models suggest.

The **observed erosion** is more consistent with a more recent formation process, which supports the idea of a **young Earth** and calls into question the chronology of hundreds of millions of years postulated by evolutionary models.

The fact that the Rockies maintain their prominence is incompatible with the amount of time proposed by evolutionary models. If these mountains had existed for tens of millions of years, they should have undergone much more extensive erosion and would not be as well preserved.

Catastrophic geological models: From the creationist perspective, catastrophic events, such as Noah's Flood and the splitting in Peleg's time, which most likely generated tectonic plate movements, could have formed these mountains and other mountain ranges in a relatively short period of time.

These geological events could have rapidly lifted the mountains, and the erosion that has occurred since then would be the result of only a few thousand years, rather than millions.

This model is more consistent with current observation of mountains, where the degree of erosion is more consistent with a young Earth than with a chronology of millions of years.

Estimated erosion rate: At erosion rates observed today, it is estimated that mountains lose several millimeters to centimeters per year, depending on conditions.

If the Rocky Mountains had existed for tens of millions of years, they should have been significantly reduced, much more than what we observe today.

This mismatch between current erosion rates and the degree of observable weathering reinforces the idea that the Rocky Mountains and other similar geological formations are much younger than conventional models suggest.

The degree of erosion observed in the Rocky Mountains does not agree with the idea that these mountains are hundreds of millions of years old.

This argument is consistent with the view of a young Earth, where the Rockies and other mountains formed rapidly as a result of catastrophic geological events rather than a slow, gradual process that takes millions of years.

Common counterarguments:

Some geologists argue that the Rocky Mountains have been rejuvenated by tectonic activity, which explains their relative height and current prominence.

However, continued erosion should have caused much greater wear on the mountains, even if there was some recent tectonic activity.

Moreover, the lack of significant erosion is difficult to explain when you consider that these mountains have existed for tens of millions of years.

Snelling, A. A. (2009). Earth's Catastrophic Past: Geology, Creation, and the Flood. Institute for Creation Research.

Morris, J. D. (1994). The Young Earth: The Real History of the Earth - Past, Present, and Future. Master Books.

Austin, S. A. (1994). Grand Canyon: Monument to Catastrophe. Institute for Creation Research.

24) Fresh Sedimentary Rocks:

Many sedimentary rock formations show features that indicate rapid deposition rather than slow accumulation over millions of years.

This argument is based on the observation that many of these formations feature horizontal and flat layers with very little erosion between them, which is more consistent with rapid deposition during catastrophic events, such as Noah's Flood, rather than the slow geological processes proposed by evolutionary models.

Evidence observed:

Horizontal and flat layers with little erosion: Horizontal layers are observed in many sedimentary rock formations that are evenly and flatly arranged, without showing the typical signs of erosion between the layers.

If these layers had been deposited slowly over millions of years, we would expect to see more signs of wear and erosion between them, due to prolonged exposure to the elements such as wind and rain.

The lack of significant erosion between the layers is a strong indication that the sediment layers were deposited rapidly, probably in the context of a catastrophic event such as a massive flood, where large amounts of sediment were deposited in a short period of time.

Fossil preservation: Many sedimentary rocks contain well-preserved fossils, suggesting that organisms were buried quickly, before they had time to decompose or be washed away by water.

Rapid burial is necessary to preserve the structure of the fossils in such detail, as decay and scavengers would destroy the remains if they were exposed for extended periods.

This type of conservation is most consistent with a catastrophic event that caused rapid deposition of large amounts of sediment, which quickly buried the organisms.

Noah's Flood, in the creationist model, is the most cited event that would explain this phenomenon.

Vast and uniform sedimentary formations: Vast and continuous sedimentary formations are observed that extend over large geographical areas, which is difficult to explain by slow and local deposition.

These formations suggest that the sediment was deposited evenly and rapidly, which is more consistent with a massive global or regional event.

In catastrophic events, such as the Flood, large amounts of sediment would have been washed away and deposited in a short period, covering vast areas and creating the flat, spread layers of sediment that we observe today.

Lack of bioturbation: Bioturbation is the alteration of sediments by the activity of living organisms, such as earthworms or crustaceans, which dig into the soil and modify the sedimentary structure.

In many sedimentary formations, a lack of bioturbation is observed, suggesting that the layers were deposited quickly and that organisms did not have time to alter the sediment.

If sedimentary layers had slowly accumulated over millions of years, one would expect to find more signs of bioturbation in them, as organisms would have plenty of time to interact with the sediment.

Implications for the age of the Earth:

The lack of significant erosion between sedimentary layers, the presence of well-preserved fossils, and the lack of bioturbation all point to these sedimentary rocks being deposited rapidly in a catastrophic event rather than slow, gradual deposition over millions of years. This type of rapid deposition is most consistent with a young Earth, where important geological processes occurred in a short period of time.

Comparison with evolutionary models:

Evolutionary models propose that sedimentary rocks formed slowly over millions of years, with sediment accumulating over time.

However, the features observed in many sedimentary formations, such as the horizontality of the layers and the preservation of fossils, are not consistent with slow and prolonged deposition.

Instead, these traits are more consistent with rapid deposition, such as would have occurred during Noah's Flood, according to the creationist model.

This catastrophic event would have rapidly deposited large amounts of sediment around the world, which explains the lack of erosion between the layers and the preservation of fossils.

Common counterarguments:

Evolutionary geologists suggest that the lack of erosion between sedimentary layers could be due to periods of geological stability or the fact that the layers were deposited underwater, which could have reduced erosion.

However, even in aquatic environments, some erosion and bioturbation would be expected if the layers had remained exposed for extended periods of time.

The features seen in many sedimentary formations, such as horizontal layers without significant erosion and the presence of well-preserved fossils, are more consistent with rapid deposition during catastrophic events than with slow accumulation over millions of years.

This supports the idea of a young Earth, where the most important geological processes occurred in a short period of time, rather than the slow gradual changes proposed by evolutionary models.

Snelling, A. A. (2009). Earth's Catastrophic Past: Geology, Creation, and the Flood. Institute for Creation Research.

Morris, J. D. (1994). The Young Earth: The Real History of the Earth - Past, Present, and Future. Master Books.

Austin, S. A. (1994). Grand Canyon: Monument to Catastrophe. Institute for Creation Research.

25) Expansion of the Sahara:

The Sahara Desert is the largest hot desert in the world and is constantly expanding southward due to desertification.

Studies of the age and expansion of the Sahara suggest that this desert is between 4,000 and 6,000 years old, which is more consistent with a young Earth model.

If the Earth really were millions of years old, as evolutionary models suggest, it would be reasonable to expect that the Sahara would be much larger or that it would have gone through multiple cycles of formation and disappearance over time.

However, the Sahara's relatively young age raises questions about models of an ancient Earth.

Evidence observed:

Sahara Expansion Rate: It has been observed that the Sahara is expanding southwards at a rate of approximately 48 kilometers per year due to desertification.

This means that the desert is covering more extensive areas as the dry weather progresses and the surrounding areas lose vegetation.

This process has been studied and monitored in recent decades, and it is estimated that the expansion of the Sahara began approximately 4,000 to 6,000 years ago, coinciding with significant geological and climatic events in Earth's history.

Estimated age of the Sahara: Studies of the age of the Sahara suggest that this desert formed between 4,000 and 6,000 years ago, after a major climate change that transformed what was once a wetter region into the desert we see today.

Evidence of ancient lakes, rivers, and increased biodiversity has been found in what is now the Sahara Desert, indicating that the region was not always arid and deserted.

These changes point to a relatively recent event in geological terms, which is not consistent with a chronology of millions of years.

Cycles of formation and disappearance: If the Earth were millions of years old, as evolutionary models suggest, the Sahara would have had to go through multiple cycles of formation and disappearance, with the desert expanding and contracting over long periods of time.

However, there is no evidence that the Sahara has existed in such long cycles.

Instead, what we observe is an ever-expanding desert, which appears to have formed recently from a geological point of view, supporting the idea of a major event that triggered its formation.

Recent climate changes: The significant climate changes that led to the formation of the Sahara appear to have occurred several thousand years ago, not millions. These changes may be related to major geological events, such as Noah's Flood, according to the creationist model.

The Flood would have caused a significant change in the global climate, and the subsequent formation of deserts such as the Sahara could be a consequence of these changes. The Sahara timeline fits well within a model that suggests a young Earth and recent geological history.

Lack of evidence of ancient cycles: Despite advances in geology and paleoclimatology, no evidence has been found that the Sahara existed as a desert for millions of years.

There is no clear evidence that the Sahara has gone through multiple cycles of expansion and contraction over millions of years, as would be expected if the Earth were as old as evolutionary models suggest.

The young age of the Sahara reinforces the idea that current geological processes began in more recent times, raising doubts about the validity of the long periods of time proposed by evolution.

Implications for the age of the Earth:

The expansion of the Sahara and its estimated age of between 4,000 and 6,000 years do not agree with a model of Earth of millions of years. If the Sahara really were millions of years old, we should observe evidence of multiple cycles of desert formation and disappearance, which is not found in the geological record. Instead, what we observe is a recently expanding desert, which fits much better with a model of a young Earth.

Comparison with evolutionary models:

Evolutionary models propose that the Earth has experienced cycles of desertification and desert formation over millions of years.

However, the evidence observed in the Sahara does not support this theory.

Rather than seeing evidence of ancient cycles, studies show that the Sahara is a relatively recent phenomenon in geological terms, with an age estimated to be only a few thousand years.

Common counterarguments:

Some evolutionary scientists argue that deserts, such as the Sahara, may have undergone changes in size over time due to climate fluctuations, but that these changes can be difficult to detect in the geological record.

However, the lack of evidence for ancient cycles of desert formation and disappearance remains a problem for models proposing a million-year-old Earth.

Griffin, D. (2002). The expansion of the Sahara Desert and its impacts on African civilizations. International Journal of Desert Studies.

Austin, S. A. (1994). Grand Canyon: Monument to Catastrophe. Institute for Creation Research.

Snelling, A. A. (2009). Earth's Catastrophic Past: Geology, Creation, and the Flood. Institute for Creation Research.

26) Colorado Canyon:

Colorado Canyon has been used as an example of gradual erosion over millions of years, but geological evidence suggests that it may have formed rapidly in a catastrophic event related to a major flood.

The Colorado Canyon has traditionally been used as a classic example of gradual erosion over millions of years.

According to the conventional view, the Colorado River would have been slowly eroding the layers of sedimentary rock over about 5 to 6 million years, giving rise to the impressive formation we see today.

The evidence for the Colorado Canyon and its rapid formation rather than a gradual process over millions of years is a topic of great interest to geologists who support the model of a young Earth.

The argument is based on three main observations:

Flat and continuous sedimentary strata: Flat and continuous sedimentary strata that extend over great distances without showing significant signs of erosion between them. This suggests that the sediment layers were deposited in rapid succession, not over millions of years.

This contradicts the notion that they formed slowly over millions of years, since, in that case, we should see signs of significant erosion between the layers.

The lack of this evidence suggests that the layers may have been deposited in rapid succession, which is more consistent with a catastrophic event.

Lack of eroded sediment: The traditional theory holds that the Colorado River slowly eroded the canyon over millions of years, but if that were true, we should find a large amount of eroded sediment in the vicinity of the canyon. However, this is not what is observed.

This supports the idea that the canyon was quickly formed by a large amount of moving water, which is consistent with a catastrophic event like the biblical Flood.

River systems: The rivers that flow into the Colorado Canyon and the formations of smaller canyons do not show the development that would be expected if the process had been gradual.

The geological structure suggests that the canyon formed in a much shorter timeframe, which is difficult to explain within the traditional evolutionary framework of millions of years.

This catastrophic interpretation of the formation of Colorado Canyon fits within the model of Noah's Flood, which could have caused the rapid displacement of large amounts of water, rapidly forming the canyon rather than over millions of years. This idea supports the notion of an Earth much younger than is usually postulated in evolutionary theory.

This approach to catastrophic erosion aligns with the account of Noah's Flood and posits that the Colorado Canyon could have been the result of large amounts of water receding during a global flood event, which would have dramatically accelerated the erosion process.

In this way, large geological formations would not require millions of years to form but could have originated in a much shorter period of time under extreme conditions, which reinforces the idea of a young Earth.

Austin, S. A. (1994). Grand Canyon: Monument to Catastrophe. Institute for Creation Research. This book examines Colorado Canyon from a catastrophic perspective, arguing that its formation may have occurred during Noah's Flood.

Morris, H. M., & Whitcomb, J. C. (1961). The Genesis Flood: The Biblical Record and Its Scientific Implications. Phillipsburg: Presbyterian & Reformed Publishing. This book pioneered the catastrophic approach to Noah's Flood, providing a scientific and biblical basis for the study of geological events.

Snelling, A. A. (2009). Earth's Catastrophic Past: Geology, Creation & the Flood. Institute for Creation Research. This work expands studies on geology in the context of a young Earth and the Flood, including discussions of the Colorado Canyon.

Oard, M. J. (2008). Flood by Design: Receding Water Shapes the Earth's Surface. Master Books. This author argues that many large geological features, including Colorado Canyon, were formed by the retreat of large amounts of water during and after the biblical Flood.

27) Basalt Pillars:

The formation of basalt pillars, also known as basaltic columns, is an impressive geological phenomenon that results from the rapid cooling of basaltic lava. When lava cools rapidly, it begins to contract and forms fractures in precise geometric patterns, usually hexagonal or polygonal.

This rapid cooling process generates basalt pillars that, in some cases, can form in days or weeks under optimal conditions, challenging the previous belief that they required millions of years to form.

Notable examples of basalt pillars include:

The Giant's Causeway (Northern Ireland): It is an iconic formation made up of more than 40,000 basalt columns.

This structure was formed during a volcanic eruption approximately 60 million years ago, according to evolutionary chronology.

However, some geologists who support the idea of a young Earth propose that such formations could have been generated in a catastrophic and rapid event, such as Noah's Flood.

Devils Postpile **(California, USA)**: It's another impressive basaltic column formation that arose from the rapid cooling of a lava flow about 100,000 years ago, according to conventional chronology.

However, these well-preserved columns suggest that, under the right conditions, this process can occur in much less time.

These basalt pillars are consistent with the possibility of rapid and violent geological events, such as those associated with the global flood mentioned in the Bible or Peleg's Division into Times.

These examples reinforce the idea that many of the geological formations do not necessarily require millions of years to form and could have occurred in much shorter times under extreme conditions.

Bibliographic References:

Austin, S. A. (1994). Grand Canyon: Monument to Catastrophe. Institute for Creation Research.

Snelling, A. A. (2009). Earth's Catastrophic Past: Geology, Creation & the Flood. Institute for Creation Research.

Oard, M. J. (2008). Flood by Design: Receding Water Shapes the Earth's Surface. Master Books.

These studies contribute to the understanding of how quickly some geological formations, such as basalt pillars, can form, supporting the idea of catastrophic events in a much shorter time frame than conventional geological models propose.

Part III: Biological Evidence

This section is not about specific organs, as I have done in previous circumstances, such as about the ear or the eye. Here we will focus on the supposed biological evidence that proponents of evolution use to support their theory.

Evolution, far from being the undeniable truth that we have been led to believe, is based on a series of assumptions that require a large dose of faith.

Often, proponents of evolution simply mock those who believe in creation, rather than presenting clear and hard evidence to support their theory.

Not only do they present their position as unquestionable, but they tend to discredit and ridicule those who raise legitimate doubts about the veracity of their claims and, *The truth does not need to ridicule others, for the truth stands on its own.*

One of the biggest problems with evolution is that it is presented as an absolute truth, with no room for debate, which is itself an anti-scientific stance.

The arguments in favor of evolution are riddled with dubious language, using expressions such as "maybe", "could be", "possibly", which indicates that these are not concrete facts but speculations.

However, when it is taught in colleges or universities, it is done without room for questioning, while belief in creation is attacked as if it were a serious error.

This is not only a problem of perspective, but of philosophical principles. Evolution is, in many ways, as faith-based a belief as biblical creation is.

Evolutionists have faith that time and chance have produced everything that exists, from DNA to the most complex ecological systems. They believe that somehow life arose spontaneously from inert matter, although this has never been directly observed anywhere.

On the other hand, creationists, who base their faith on the Bible, also find in science numerous evidence of intelligent design in nature.

The idea of a young Earth is not a simple irrational belief, but a position that finds support in observable facts and, above all, in divine principles.

Those who ridicule creation as an absurd or uncultured belief show, in many cases, an attitude of fanaticism, which closes itself to dialogue.

Evolution needs faith as much as creation.

It requires belief in processes that have never been observed, such as the transformation of one species into a different one, or the appearance of life out of nowhere.

In the end, each person must decide where to place their faith: in blind chance and infinite time, or in an intentional and divine design that leaves evident traces in creation itself.

Truth, both in science and in faith, stands on its own, and those who seek it, find it.

28) Genetic Mutations:

This argument presented in evidence about genetic mutations focuses on how mutations in each generation's DNA accumulate over time.

Below, I explain it in a simple way so that it is understandable by both people without knowledge of genetics and atheists:

Explanation with apples:

Imagine that our DNA is like a basket full of apples. Each apple represents a piece of genetic information, which is what makes us the way we are (our traits, our health, etc.).

Now, every time one generation passes its basket of apples to the next, a rotten apple appears. That rotten apple symbolizes a mutation, a small error in our DNA.

If we continue to pass the basket from generation to generation, more and more rotten apples are added. And if after many, many generations we keep adding rotten apples, in the end, If we keep adding rotten apples (mutations) to the basket in each generation, there will come a time when the basket will be full of rotten apples, which will seriously affect the functioning of our body.

That is, DNA would be so full of errors that our organism would no longer be able to function properly, and ultimately, life would be unviable.

This can be compared to trying to build a house using broken bricks. At first, some bricks may not affect as much, but if we keep adding broken bricks with each generation, the house would eventually collapse.

Relationship to evolution:

If humanity had been around for millions of years, our apple baskets would already be too full of rotten apples. This means that, if mutations accumulated over millions of years, as evolutionary theory suggests, our species would have already disappeared due to the number of errors accumulated in our DNA. Or currently we would be full of errors, all deformed and so on.

Another simple example:

It is as if each generation receives a book with a few messy pages (errors in DNA). If this process continues for millions of years, eventually the book would be so confusing that you would not be able to understand or use it.

The same goes for DNA: if humans had been around for millions of years, our DNA would be so damaged by mutations that it would be incompatible with life.

In short, this argument from the accumulation of mutations suggests that humanity has not existed for millions of years, but for a much shorter period. This casts doubt on the idea that life has existed for as long as the theory of evolution proposes, and instead supports a more recent view of life, compatible with a young Earth.

But why are we still alive? Because all is not lost! Our bodies have ways of fixing some of these rotten apples.

It's like we have a team of repairmen trying to fix the broken bricks in our house.

However, this repair team can't always fix everything. And, in addition, over time, more and more rotten apples accumulate, and the repair team has more and more work.

In other words: DNA, which contains the genetic information of all living beings, undergoes small alterations or "mutations" in each generation.

This gives us a very simple and forceful explanation of why humanity cannot be as old as some people say. With this alone, it is conclusive that Chance and Mother Nature, together with the god Time, have been able to create everything that exists. It is that instead of order we have a disorder.

Counterargument (from the evolutionary perspective):

From the point of view of evolutionary advocates, genetic mutations are not always harmful or inevitably lead to extinction.

Evolutionists argue that while it is true that many mutations are harmful or neutral, some mutations are beneficial and can be favored by natural selection, improving a species' ability to adapt to its environment.

This is crucial for evolution, since beneficial mutations are what allow the emergence of new characteristics that can improve an organism's chances of survival.

Mutation Correction and Repair Mechanisms: In addition, evolutionists argue that organisms have evolved genetic repair mechanisms, which can correct many of the errors that occur during DNA replication. *But they have not shown it.*

Cells have control systems that detect and repair DNA damage before these errors can be passed on to the next generation.

While not all errors are corrected, these mechanisms significantly reduce the rate of accumulation of harmful mutations.

Evolutionary Balance: Another aspect that evolutionary advocates mention is the balance between the rate of mutation and natural selection. Although mutations can accumulate, many of the negative or harmful mutations are quickly eliminated from the population through natural selection, i.e., individuals that carry those mutations tend to be less successful in reproduction.

In this way, evolution maintains a balance in which beneficial mutations can accumulate over time, while harmful mutations are selected against.

Neutral Mutations: In addition, there is the neutralist theory of evolution, which suggests that many mutations do not have a significant effect on an organism's survival or reproduction.

These neutral mutations can accumulate without causing major damage, as they do not directly affect the body's vital functions.

In this perspective, the accumulation of mutations does not represent a lethal burden for species, since not all mutations have negative consequences.

Evolutionary Conclusion: Thus, evolutionists' argument is that genetic mutations, along with natural selection and DNA repair mechanisms, allow species to adapt, evolve, and survive for long periods of time without cumulative mutations resulting in extinction.

Evolution is seen as a balanced process in which mutations do not inevitably lead to the destruction of a species but play a crucial role in the diversification and adaptation of life on Earth.

This counterargument is central to the debate, as it offers an alternative explanation for why mutations do not necessarily lead to extinction over a million-year frame, as evolutionary models suggest.

Errors in evolutionary reasoning:

Accumulation of harmful mutations: Although evolutionists argue that beneficial mutations and genetic repair mechanisms can compensate for harmful mutations, the reality is that most mutations are neutral or harmful, not beneficial. Beneficial mutations are extremely rare.

As time passes, harmful mutations continue to accumulate in the genome, increasing the genetic load in each generation. This accumulation of harmful mutations is not eliminated effectively enough by natural selection, eventually leading to a general deterioration in the population.

The idea that genetic repair mechanisms can compensate for all these mutations is optimistic, but inaccurate; Repair systems have limitations, and not all mutations can be corrected.

Limited role of beneficial mutations: The evolutionary approach assumes that beneficial mutations can compensate for the damage caused by harmful mutations. However, the problem is that beneficial mutations are very rare compared to harmful ones.

The probability that a mutation will benefit the organism and be positively selected is extremely low, while most mutations tend to be neutral or harmful. Therefore, the accumulation of harmful mutations over time is much more likely than the emergence of beneficial mutations that can counteract them.

There have been calculations about the likelihood that a mutation is beneficial. One of the geneticists who has addressed this issue is John Sanford, who in his book *Genetic Entropy & the Mystery of the Genome* explains that most mutations are harmful, and beneficial mutations are extremely rare. Sanford estimates that beneficial mutations are in the range of 1 in 1 million or even less common, meaning that for every beneficial mutation, there are many more that are neutral or harmful.

Another influential study in this field was conducted by H. J. Muller, a Nobel Prize winner, who in his article on *"Our Load of Mutations" (1950)* also argued that the vast majority of mutations are slightly harmful, implying that the number of beneficial mutations is negligible in comparison.

In general, many biologists (scientists, not only scientists who are on the side of evolution) agree that beneficial mutations, which are necessary for evolution to be a cumulative process that increases the complexity and functionality of organisms, are extremely rare. The probability is in the range of 1 in 10,000 to 1 in 1,000,000, although it can vary depending on the organism and environment.

And one more is, Michael Behe, in his work *The Edge of Evolution*, also mentions that even when beneficial mutations are observed, they are usually of small magnitude and, many times, imply the loss or degradation of some pre-existing genetic function, which casts doubt on their contribution to an increase in biological complexity.

The conclusion of the studies is that the accumulation of harmful mutations over time is inevitable, and the rate of occurrence of beneficial mutations is too low to balance the growing "mutational burden."

Insufficient natural selection: Although natural selection removes some harmful mutations, it cannot eliminate all of the harmful mutations that accumulate in the genome.

In fact, many slightly harmful mutations can go undetected by natural selection and accumulate over generations, gradually weakening the population.

This process is known as "genetic degeneration" and suggests that natural selection is not a foolproof mechanism for maintaining genetic stability over millions of years.

Neutral mutations and degeneration: The argument for neutral mutations, while valid, also has its flaws.

Many seemingly neutral mutations can have long-term negative effects, subtly affecting the genome and contributing to the mutational load.

In addition, neutral mutations that accumulate in non-coding regions of DNA can lead to genetic deterioration if at any time they affect gene regulation or control of essential genes.

Insufficient time for genetic balance: The argument that beneficial mutations compensate for harmful mutations ignores the fact that the time needed for beneficial mutations to become fixed in the population is very long compared to the rapid accumulation of harmful mutations.

The mutation rate and cumulative effects of harmful mutations far outweigh the ability of beneficial mutations to fix or balance the genome, especially when considering time scales of millions of years.

In short, the accumulation of harmful mutations over generations, coupled with the low frequency of beneficial mutations, refutes the argument that evolution by natural selection can keep populations genetically healthy over millions of years. This process of genetic degeneration is incompatible with the long chronology proposed by evolutionists and points to a more recent history of humanity, which is consistent with a young Earth.

Some key references:

Sanford, J.C. (2008). *Genetic Entropy & the Mystery of the Genome. FMS Publications.* Sanford, a geneticist, makes detailed arguments about how negative mutations tend to accumulate over time, weakening populations. Their work is fundamental to understanding the concept of genetic degeneration and how harmful mutations outweigh beneficial ones.

Behe, M.J. (2007). *The Edge of Evolution: The Search for the Limits of Darwinism. Free Press.*

Muller, H.J. (1950). *Our Load of Mutations. American Journal of Human Genetics, 2(2), 111-176.* This landmark paper by Muller, a renowned geneticist, introduces the concept of "mutational load," which refers to the accumulation of genetically harmful mutations over time.

Lynch, M. (2010). *Rate, molecular spectrum, and consequences of human mutation. Proceedings of the National Academy of Sciences, 107(3), 961-968.* In this article, Lynch discusses the rate of mutations in humans and the impact they have on genetic evolution, highlighting how harmful mutations can accumulate faster than beneficial mutations can eliminate them.

Kondrashov, A. S. (1995). Contamination of the genome by very slightly deleterious mutations: Why have we not died 100 times over? Journal of Theoretical Biology, 175(4), 583-594. Kondrashov addresses the accumulation of mildly detrimental mutations in populations and raises questions about why populations have not gone extinct due to this accumulation.

These sources provide a solid scientific basis on the debate of genetic mutations and their implications in evolutionary times.

29) Extinction of Species:

The argument about species extinction is based on the fact that, if the Earth really were millions of years old, we should observe a much higher extinction rate relative to today's biodiversity. Here is the most developed argument:

Extinctions and Current Biodiversity: If the Earth were extremely old (millions of years), it would be logical to assume that many more species would have gone extinct over time.

This is similar to the toy box metaphor: if that box has been around for a long time, many of its "toys" (species) should have disappeared or been deteriorated.

However, the amount of biodiversity present today seems incompatible with millions of years of history. If the rate of extinction had been steady or even faster on such an ancient planet, we should find many more extinct species and less biodiversity.

This observation is used by some young Earth advocates as a sign that time is not as long as it is proposed in the evolutionary view.

Fossil Record and Catastrophic Events: The fossil record also does not support a gradual extinction over millions of years. On the contrary, it shows that many species disappeared abruptly due to catastrophic events, such as asteroid impacts or large volcanic eruptions.

These catastrophic events would be more consistent with a more recent history of the Earth, rather than one that spans millions of years.

These kinds of sudden extinctions, observed in geological strata, reinforce the idea that the Earth changes could have occurred in a much shorter time frame.

Evolutionary Challenges: Evolution by natural selection requires a slow and gradual process over millions of years. However, the existence of these catastrophic events calls into question whether species have had enough time to adapt slowly. This could suggest that extinctions and the emergence of new species occurred in a much shorter span of time than traditional evolutionary theory suggests.

For the literature on the topic of species extinction and the arguments for a young Earth, the following general sources can be included:

Morris, Henry M. The Genesis Flood: The Biblical Record and Its Scientific Implications. This book addresses catastrophic events such as Noah's Flood and its impact on species extinction, questioning the narrative of gradual evolution and Earth's long history.

Snelling, Andrew A. Earth's Catastrophic Past: Geology, Creation & the Flood. This author focuses on the geological and paleontological aspects that support a biblical interpretation of the Earth's history, including the issue of species extinction in relation to the young Earth.

Whitcomb, John C., and Henry M. Morris. The Genesis Flood. It is one of the classic texts in the discussion of creation and the Flood, arguing that catastrophic events shaped the Earth rapidly, which affects the interpretation of the fossil record.

Austin, Steven A. Grand Canyon: Monument to Catastrophe. A geological analysis of the Grand Canyon, arguing that formations like these and associated extinctions fit a catastrophic model, compatible with a young Earth.

Woodmorappe, John. Noah's Ark: A Feasibility Study. This study offers insight into how species extinctions might be related to a global flood and a biblical time frame.

These texts provide a scientific and theological background that argues for one-off catastrophic events and mass extinction in a more recent time frame, supporting the idea of a young Earth. Although it is a pity that the event that occurred in Peleg's time there is so little info, and we do not enter into assumptions like unbelievers.

30) Biological Clocks

Dendrochronology

Biological clocks are internal mechanisms *in living organisms* that regulate various physiological processes as a function of time. These clocks are crucial for keeping pace and timing biological functions, such as growth, aging, reproduction, and sleep cycles.

Dendrochronology is the study of the growth rings in trees. Each year, trees produce a new growth ring, and these rings can be analyzed to gain insight into the age of the tree and the environmental conditions it experienced during its lifetime.

Biological clocks in cells, such as telomeres, suggest a limited lifespan.

Telomeres are primarily composed of **repetitive DNA** and a set of **specific proteins** that work together to protect the ends of chromosomes.

Telomeres are made up of repeats of a short DNA sequence rich in the basis's guanine and thymine (in humans, this sequence is repeated thousands of times as TTAGGG).

Telomeres are associated with several specific proteins, such as shelterin complex proteins, that stabilize and protect the end of the chromosome. These proteins include TRF1, TRF2, POT1, among others, and help prevent the cell's DNA repair system from mistaking the telomere for a damaged DNA fragment.

L-Glutamine and L-Lysine: These amino acids are important for protein synthesis and maintenance and can support the stability of the shelterin complex. And there are other place recommendations to prevent telomeres from shortening.

Telomeres, which protect the DNA in chromosomes, shorten with each cell division, a process that could not be sustained for millions of years in complex organisms.

Imagine a shoelace with plastic on each end that protects it from fraying; Every time you tie it, that plastic wears out.

Similarly, whenever a cell divides, telomeres shorten to a point where the cell can no longer divide and dies.

This implies a biological time limit for life on Earth.

If life had existed for billions of years, ancient organisms would have had to go through so many cell divisions that their telomeres would be exhausted, which is not the case.

Telomeres, then, impose a limit on cellular longevity and, therefore, on life in complex organisms.

In a scenario of millions of years, we should find organisms with extremely short telomeres or without them, which does not happen.

Other Relevant Biological Clocks are:

1. **DNA methylation**: This process adds methyl groups to certain DNA bases and changes with age, indicating a "biological age."

2. **Histone modifications**: Histones, proteins around which DNA is coiled, lose effectiveness over time. I have not included this one and within the 81 many more came to light,

   I have chosen to structure 81 evidences for the symbolism of the number, but, in reality, the evidences could be vast, perhaps countless. This focus limited in number, but broad in depth, gives the project a manageable structure, while suggesting that there is much more evidence that could support a young Earth and universe.

3. **Chaperone proteins**: These proteins help the folding of other proteins; over time, their efficiency decreases, contributing to neurodegenerative diseases.

4. **Microbiome**: The community of microorganisms in our body changes with age, influencing health and longevity.

5. **Cell cycle**: A process regulated by proteins that control cell division, with errors that increase aging and disease risk.

6. **DNA damage**: Cells accumulate DNA damage from factors such as radiation and chemicals, leading to aging or cell death.

7. **Free radicals**: Unstable molecules produced by metabolism that, when accumulated, damage cells and accelerate aging.

These biological mechanisms act like internal clocks, and if organisms had existed for millions of years, these processes would have failed, pointing toward a much shorter biological limit for life on Earth and thus supporting the idea of a young Earth.

Dendrochronology: The World's Oldest Tree

The oldest recorded tree, known as "Methuselah," a long-lived pine (Pinus longeval), is approximately 4,846 years old, an age that coincides with the time after the biblical Flood (according to biblical chronology it happened about 4,400 years ago).

This tree is located in the White Mountains of California, USA, in an undisclosed location to protect it. Its age is determined by carotage or core extraction, using a Pressler auger to extract a wooden cylinder, but without damaging the tree and counting its growth rings, each representing a year. The tree is about 15 meters tall.

Methuselah is considered a "living witness" to historical events, including climate and environmental changes over the past millennia.

Its longevity challenges the idea of great cycles of time in Earth's history and aligns best with a chronology of thousands, not millions of years.

Long-lived pines, such as Methuselah, survive in arid and harsh conditions, at high altitudes with little soil and water, factors that contribute to their longevity.

If the Earth really did have conditions suitable for life for hundreds of thousands or millions of years, one would expect to find long-lived trees or organisms that reflected those ages.

Examples such as the long-lived pine or giant sequoias have the capacity to live thousands of years, but the longest-lived trees we know, such as Methuselah, reach a maximum age of close to 4,800 years.

In a million-year scenario, it would be reasonable to find trees that are 8,000, 15,000, or 25,000 years old, which begs the question: Why aren't there significantly older trees? This suggests that the Earth has not been in its current state for millions of years.

I do not intend to extrapolate a definitive conclusion about the age of the Earth from a single case, but it is also unreasonable to ignore its contribution to the body of evidence.

Although evolution tends to draw conclusions from a single fossil or bone fragment, this work seeks to add up each piece of evidence as part of a larger set.

The Bible states, categorically and unambiguously, that "In the beginning God created the heavens and the earth," a direct message that, unlike the ambiguous terms of evolution, does not depend on assumptions or possibilities.

Some references to support these issues:

Elizabeth H. Blackburn, Carol W. Greider, and Jack W. Szostak: Their work on telomeres and cellular longevity, ***recognized with the Nobel Prize in 2009,*** is fundamental. You can find his research in some of his key papers published in high-impact journals, especially in **Cell** and **Annual Review of Biochemistry**.

Blackburn, E. H., Greider, C. W., & Szostak, J. W. "Telomeres and Telomerase: The Pathway to Immortality." Cell, y Annual Review of Biochemistry.

A. Olovnikov: Theory that telomere shortening limits cell replication, published in "A Theory of Marginotomy" in the Journal of Theoretical Biology (1973).

Edmund Schulman: Pioneer in studies of the longevity of long-lived pines in the White Mountains, with the identification of the "Methuselah" tree.

Schulman, E. "Longevity under Adversity in Conifers." Science (1958).

Donald R. Currey: Research in dendrochronology, including dating of growth rings in ancient trees.

Currey, D. R. "Dating Tree Rings of Pinus Longaeva." American Journal of Botany (1965).

Henry M. Morris. The Genesis Flood: The Biblical Record and Its Scientific Implications. He relates the events of the Flood to the interpretation of a young Earth and rapid extinctions and a pity that he did not deal with The Splitting of La Pange in Peleg's Time.

Andrew A. Snelling. Earth's Catastrophic Past: Geology, Creation & the Flood. It examines the geological and biological implications of the Flood on species extinction, reinforcing biblical chronology.

They can access them in academic databases such as **PubMed**, **ScienceDirect** or **JSTOR**, or through university libraries that subscribe to these scientific journals.

31) Soft Tissue Fossils:

The discovery of fossils containing **soft tissues**, such as muscles, blood vessels, and even cellular structures, poses serious challenges to the idea that these fossils are millions of years old.

Under normal conditions, soft tissues decompose rapidly after death, as they are highly vulnerable to putrefaction and disintegration processes. Even in highly favorable environments, the chance of these tissues lasting for millions of years is extremely low.

Under the **most favorable conditions** (e.g., cold, dry, or oxygen-free environments), soft tissues can be preserved for thousands of years, but not millions. The **preservation of soft tissues in fossils millions of years old** is surprising because, even under optimal conditions, the stability of organic molecules usually does not exceed a few thousand years before they decompose. Here I explain in detail:

Duration in favorable conditions

1. **Mummies and Dry or Frozen Preservation**: In extreme conditions such as freezing or arid environments, soft tissues can be preserved **for up to a few thousand years** (as is the case with human mummies or mammoths preserved in ice). For example:

    o **Egyptian Mummies**: Preserved in dry climates, they can be between **3,000 and 4,000 years** old.

    o **Natural Mummies on Ice**: Such as the famous "Iceman" (Ötzi), which is estimated to be around **5,300 years old**.

2. **Oxygen-Free Conditions and Rapid Burial**: When an organism becomes trapped in places without oxygen and quickly covered by sediment, some tissues may be better preserved due to lack of oxygen and bacterial inactivity. However, even in these environments, soft tissue preservation rarely exceeds **tens of thousands of years**.

3. **Organic Molecules and Proteins**: Scientific studies have shown that proteins and other organic compounds in soft tissues tend to degrade irreversibly over time, even in the absence of factors that accelerate breakdown (such as oxygen, moisture, or bacterial activity). Molecular breakdown and the breaking of bonds in proteins and DNA would generally not allow tissue preservation for millions of years.

The Difference with Million-Year-Old Fossils

What makes the soft tissue findings in dinosaur fossils millions of years old surprising is that, based on what we know about tissue chemistry, these organic materials **should have completely disintegrated** in a much shorter timeframe.

The presence of soft tissue in such old fossils contradicts the pattern of decay we observed in other remains of much younger age and suggests that these tissues have been fossilized under exceptional conditions, raising questions about the chronology and processes of fossilization.

The preservation of soft tissues in fossils suggests that these remains may be much younger than traditionally believed.

Imagine a **sandwich** left outdoors: the bread dries out, the cheese deteriorates, and the ingredients begin to decompose quickly.

Similarly, when an organism dies, flesh, skin, and other soft tissues break down rapidly.

Thus, when an organism is fossilized, it is expected that only the hardest parts, such as bones or teeth, will remain intact.

However, finding **dinosaur fossils with well-preserved soft tissues** is comparable to finding a sandwich in perfect condition after thousands or millions of years.

This phenomenon raises important questions:

- If these fossils really are millions of years old, how did the soft tissues survive?

- What factors could explain this preservation, and why do we not find them in other equally ancient fossils?

32) Ancient Bacteria:

Argument: The reactivation of bacteria found in amber and ice that have lain dormant for supposed millions of years suggests that Earth might be younger than generally claimed.

These "dormant" bacteria have come back to life when placed in the right conditions, which raises questions, as DNA and cells would be expected to disintegrate over such a long period. The ability to survive intact co*ntradicts the notion of millions of years.

Nobody thought about that in the movie Jurassic Park where it is stated that: a mosquito bites a dinosaur, then this mosquito is trapped in a Dominican amber, (although amber is not exclusive to the Dominican Republic, although Dominican amber is famous for its high quality and the quantity of fossils it preserves. Other important amber deposits are found in Mexico, *Baltic Sea regions* (as in Russia and Poland), Myanmar (Burma), and the United States (especially in New Jersey). Each of these deposits has different characteristics and ages, reflecting the geological conditions of their regions) then they take the DNA of those dinosaurs that bit those mosquitoes (they must have been super mosquitoes to penetrate the supposed skin of those dinosaurs) and with the DNA they "create" the same species that the mosquito had sucked their blood.

The point is that, with the story, many were immediately indoctrinated, through the use of cinema.

The idea popularized dinosaur genetics, but it poses serious inconsistencies, especially with an extinction that occurred about 66 million years ago.

Some additional inconsistencies in *Jurassic Park* that defy the principles of evolution and biology include:

1. **DNA conservation**: DNA cannot last for millions of years. Even under extreme conditions, DNA breaks down gradually, and it would be unlikely to find complete, usable DNA after 66 million years.

2. **DNA compatibility**: Even if fragmented DNA were obtained, its assembly and manipulation to recreate a dinosaur would be highly speculative, as we lack a "complete map" to interpret and reconstruct dinosaurian DNA.

3. **Genetic Limitation and Cloning**: Extracting and mixing DNA with current species would not guarantee the exact reconstruction of the extinct species, since genes, in species separated by millions of years of evolution, would not fit easily.

4. **Habitat and Life Cycle**: Even if a dinosaur were to be "revived," there would be challenges in adapting it to a modern environment, raising questions about its life cycle and adaptation in our current ecosystem.

These points underscore the scientific and logical difficulty of recreating extinct species from ancient DNA remains.

Comparison: It's like finding fresh food after millions of years in a fridge, which doesn't square with our understanding of life and decay.

These findings have sparked debates about long-term survivability and suggest that the estimated geologic time might require revision.

Scientific Support and Sources

Vreeland, R. H., Rosenzweig, W. D., & Powers, D. W. *(2000). Isolation of a 250-million-year-old halotolerant bacterium from a primary salt crystal.* ***Nature****,* 407(6806), 897-900. This study describes the reactivation of a bacterium found in salt crystals supposedly millions of years old.

Cano, R. J., & Borucki, M. K. *(1995). Revival and Identification of Bacterial Spores in 25- to 40-Million-Year-Old Dominican Amber.* **Science**, *268(5213), 1060-1064. It analyzes bacteria in amber that were "revived" after millions of years.*

33) Mitochondrial DNA Analysis:

Argument: Mitochondrial DNA has a faster mutation rate than initially thought, suggesting that humanity could have emerged only a few thousand years ago.

Imagine your body as a factory. Inside each cell of that factory are tiny powerhouses called mitochondria. These mitochondria have their own DNA, much smaller than the DNA we have in the nucleus of our cells. This special DNA is called mitochondrial DNA.

An important feature of mitochondrial DNA is that it is inherited exclusively from the mother. That is, you receive your mitochondrial DNA from your mother, she from her mother, and so on. Inherited only through the maternal line, mitochondrial DNA allows us to trace our maternal ancestry through generations.

Like a clock, mitochondrial DNA accumulates mutations (changes in its sequence) over time. The more time passes, the more mutations there will be. Therefore, mitochondrial DNA can be used as a "molecular clock" to estimate when two populations separated or when a species emerged.

If the mutation rate is faster, then genetic differences between human populations could have accumulated in a shorter period of time. This could suggest that humanity emerged less time ago than traditionally thought.

If it was not understood, let's see in other words, but we are all descended from EVA. THAT IS THE CONCLUSION. WHETHER THEY LIKE IT OR NOT. AND THE BIBLE PROVES THAT IT IS NOT MISOGYNISTIC.

Mitochondrial DNA (mtDNA) is a special part of our DNA found in the mitochondria, the "energy factories" of our cells. This mtDNA is passed down only from mother to child, making it a very useful tool for tracing family lineages and understanding the origins of humanity.

More recent research has shown that the mutation rate of mtDNA is actually faster than previously thought. This changes things. If mutations occur faster, it means that humanity's "age," based on mtDNA, is much younger than previously believed. Some studies suggest that we could have emerged only a few thousand years ago, which fits much more in line with biblical chronology than with the idea of a humanity that is hundreds of thousands of years old.

If we follow the mtDNA trail of all humanity, we come to a woman who lived a few thousand years ago, not hundreds of thousands ago, which is much more in keeping with the narrative of a young humanity, as described in the Bible.

Some key studies and references include:

***Cann, Rebecca L., Stoneking, Mark, and Wilson, Allan C.** (1987). Mitochondrial DNA and human evolution. **Nature**, 325(6099), 31-36.* This study introduced the concept of "mitochondrial Eve," suggesting that all of humanity shares a common maternal line.

***Parsons, T.J., et al.** (1997). A high observed substitution rate in the human mitochondrial DNA control region. **Nature Genetics**, 15(4), 363-368.* This research found a higher-than-expected mutation rate in mitochondrial DNA, which could indicate a shorter timeline for humanity.

***Gibbons, A.** (1998). Calibrating the Mitochondrial Clock. **Science**, 279(5347), 28-29.* This article discusses the debate over the mutation rate of mitochondrial DNA and its implications for human chronology.

These studies and articles are relevant to exploring how the mutation rate of mitochondrial DNA may influence estimates of humanity's antiquity.

34) Ecosystem Collapse:

Argument: The stability of entire ecosystems is unsustainable over millions of years due to the fragility of food chains and biological interdependencies.

Ecosystems, with their intricate webs of life and interdependent relationships, are inherently fragile and cannot maintain their stability over millions of years. Each species in an ecosystem depends on others to survive.

If a species disappears, it can set off a chain reaction that affects the entire ecosystem. Events such as volcanic eruptions, sudden climate changes, or diseases can decimate populations and disrupt food chains.

Ecosystems are complex, made up of many species that depend on each other for survival. For example, predators need prey to feed on, and plants need pollinators to reproduce. Everything is interconnected, like in a food chain. Now, if this delicate balance of ecosystems had lasted for millions of years, small disturbances in those relationships could have caused major collapses.

A small change, such as the extinction of a keystone species (such as pollinators or prey), could destabilize the entire ecosystem. If species were living together for millions of years, the odds of these collapses happening would be very high.

Instead, what we observe is that many ecosystems have remained relatively stable for long periods, which is more consistent with a young Earth scenario, where these ecosystems have not had to endure the fragility of interdependence for millions of years. In short, the stability of ecosystems is difficult to sustain for so long, suggesting that they have not existed for millions of years, but for thousands.

Some references to the fragility and stability of ecosystems and their difficulty in sustaining themselves for millions of years:

Odum, Eugene P. *Fundamentals of Ecology.* This classic ecology book discusses the interdependence in food chains and the instability that occurs when a keystone species is altered.

Terborgh, John, et al. (2001). *Ecological Meltdown in Predator-Free Forest Fragments.* **Science**, 294(5548), 1923-1926. This study highlights how predator removal destabilizes ecosystems and shows the fragility of ecological interdependencies.

Pimm, S. L. (1991). *The Balance of Nature? Ecological Issues in the Conservation of Species and Communities.* **University of Chicago Press**. It examines how the extinction of a species can cause ecological collapses, arguing that ecosystems have limits to long-term stability.

These sources present evidence that long-term stability in complex ecosystems is limited, raising questions about the ability of ecosystems to sustain themselves over millions of years.

35) Animal Survival Record:

Plot: Current species show genetic characteristics that would be unsustainable if they were millions of years old. For example, the genetic complexity of certain animals does not agree with a gradual and sustained evolution over millions of years.

When we look at animals today, we find that they have an impressive genetic complexity. This complexity refers to how their genetic information is organized, which determines things like how they grow, how they reproduce, and how their bodies function.

The theory of evolution suggests that this complexity developed slowly over millions of years, through small changes or mutations in the animals' DNA. However, there are species with such complex genetic characteristics that it does not seem possible or likely that these slow, gradual changes would have been sufficient to produce such a level of sophistication in their genetics.

In other words, if species had existed for millions of years and evolved slowly, we wouldn't have expected to see so much complexity in their DNA. It's as if such a complicated system couldn't have been formed in small steps but seems to have been designed with all that complexity from the start.

This casts doubt on the idea that everything developed over millions of years in a gradual fashion and suggests that current species have not been here for that long.

The theory of evolution, as it stands today, demands a considerable burden of faith in processes that have never been directly observed and on timescales that are difficult to reconcile with the biological evidence we have around us. Data on genetic mutations, the fragility of ecosystems, and biological clocks such as telomeres and the Hayflick limit, as well as the surprising discoveries of soft tissues in fossils and preserved bacteria, lead us to seriously question the idea of an Earth millions of years old.

What we observe in modern biology suggests that life has not had millions of years to accumulate mutations or to maintain stable ecosystems. Instead, species appear to have emerged recently, in historical terms, and show a complexity that defies the gradualist explanation of evolution. These biological tests reinforce the possibility of a young Earth, where the processes of life have not been occurring for millions of years, but for a much shorter period, as described in the Bible.

Just as you can't build a house out of broken bricks, you can't expect life on Earth to have evolved over millions of years without crumbling under the weight of mutations, extinctions, and fragile ecosystems. Biological science, rather than refuting, seems to confirm a more recent history consistent with divine creation.

If the reader wishes to support the argument about genetic complexity and sustainability over millions of years in animals, here are some relevant sources:

Behe, Michael J. Darwin's Black Box: The Biochemical Challenge to Evolution. Behe discusses the "irreducible complexity" in biology, suggesting that many biological structures are too complex to have formed gradually.

Sanford, J. C. Genetic Entropy & the Mystery of the Genome. It examines how cumulative mutations negatively affect genetic information in organisms, challenging the idea of prolonged, positive evolution.

Limitations of the Hayflick Limit: The limit of cell divisions of eukaryotic cells, known as the Hayflick limit, establishes a maximum of cell divisions before senescence, which implies limitations for the longevity of multicellular organisms.

Part IV: Physical and Atmospheric Evidence

36) Atmospheric Helium and Young Earth

Within the Physical and Atmospheric Evidence is in the first place Helium which, because it is so extensive, as it already happened with the case of overpopulation and remains of these that should exist and also with the case of the ear, the eye, and others, this case due to its extension I share it in a solitary way.

Helium, the second-lightest gas after hydrogen, plays a crucial role in challenging assumptions of a billion-year-old Earth, the fundamental basis of evolutionary theory. Below are several key points that demonstrate how helium contributes to the refutation of evolutionary chronology:

Current levels of helium in Earth's atmosphere are inconsistent with an age of billions of years. Helium is generated under or below the Earth's surface and slowly escapes into the atmosphere. If Earth were really billions of years old, we should find a lot more helium accumulated in the atmosphere, as this gas escapes very slowly into outer space.

Dr. Larry Vardiman, in his comprehensive study entitled *The Age of the Earth's Atmosphere* (1990), showed that, under current rates of helium release into the atmosphere, helium accumulation would have taken less than 2 million years. While this figure may be higher than some creationists believe, it is still drastically lower than the 4.6 billion years evolutionists claim for Earth.

Factors considered by evolutionary scientists:

• The steady rate of helium escaping into the atmosphere, assuming it hasn't changed over the years.

• The premise that when Earth formed it did not contain helium in the atmosphere, starting from scratch. A logical deduction, I think.

This discovered data reduces the supposed age of the Earth by 96.96%, simply by observing helium, leaving evolutionists without the time they need to sustain their faith in an Earth of billions of years.

37) Terrestrial Helium: A Mystery

Helium is also found in large quantities beneath the Earth's crust, particularly in the zircon crystals present in granite. However, rocks are not able to retain helium indefinitely due to their light nature, and helium should have escaped long ago if Earth were as old as evolutionists claim.

Despite this, large amounts of helium have been found trapped in zircon crystals, suggesting that these rocks have not had millions of years to release helium and that they still have helium. This finding points to a recent formation of granite and, by extension, of the Earth itself.

This implies that, if the Earth really was billions of years old, helium would have dissipated by now, which contradicts current observations.

38) Helium Leakage and Radioactivity in Rocks

The natural process of radioactive decay in rocks produces helium, but this process cannot explain the enormous amount of helium trapped beneath the Earth's crust. If this process had been going on for billions of years, we should find less helium in the Earth's interior and much more in the atmosphere.

This suggests that the time available for helium escape has been much shorter than evolutionary models propose, providing another strong indication that Earth is young.

39) The Evidence for Helium in Polar Ice

Another argument related to time and helium comes from studies of ice cores extracted from Greenland and Antarctica. These ice sheets are interpreted as indicators of hundreds of thousands of years of climate history. However, the discovery of a squadron of World War II aircraft buried under more than 250 feet of ice in just a few decades suggests that ice sheets form much faster than previously thought.

If ice sheets form at a much faster rate, then the time needed to accumulate current levels of helium and other substances in the atmosphere is much shorter, again supporting a more recent chronology.

40) The Powerful Force of Helium in the Universe

In addition to its terrestrial implications, helium also plays a key role in fine-tuning the universe, an argument that assumes an intelligent creator and is rejected despite being based solely on science. There is a powerful force that holds atoms together and makes possible the existence of elements such as hydrogen, helium, oxygen, among others.

Fine Adjustment Details:

- If this force were only 5% weaker, the only element in the universe would be hydrogen, making life impossible.

- If it were only 5% stronger, the atoms would clump together into giant molecules, making life equally impossible.

This delicate balance is further testimony to the improbability that life and the universe have formed by mere chance over billions of years, as evolution maintains.

Helium, both in the atmosphere and in the Earth's interior, provides powerful evidence that challenges assumptions of a billion-year-old Earth. From helium levels in the atmosphere to its entrapment in zircon crystals, the data indicate that Earth is much younger than evolutionists propose. Models that rely on long periods of time simply cannot explain the current presence of helium.

Helium is strong evidence that the dictatorship of time in the evolutionary narrative is far from an undeniable fact, and that the evidence points to a young, precisely engineered Earth. There are flaws, errors, assumptions, as I have explained in other articles on the dates and dating methods that evolutionists have adopted to defend their philosophical creationist belief and therefore their faith. It is not true that there is overwhelming scientific evidence, quite the contrary, there is evidence in the direction in which it points to a young earth.

Scientific sources and references that support the arguments about helium in the atmosphere, the interior of the Earth, and the fine-tuning of the universe:

Vardiman, L. (1990). The Age of the Earth's Atmosphere: A Study of the Helium Flux through the Atmosphere. This study looks at helium accumulation rates and how this challenges a billion-year-old Earth.

Humphreys, D. R. (2005). Young Helium Diffusion Age of Zircons Supports Accelerated Nuclear Decay. RATE II. This paper explores helium accumulation in zircon crystals, suggesting shorter retention times.

Gonzalez, G., & Richards, J. W. (2004). The Privileged Planet: How Our Place in the Cosmos Is Designed for Discovery. This book introduces the idea of fine-tuning, explaining how the properties of elements such as helium support an intentional design of the universe.

These references offer a scientific basis for considering the possibility of a young Earth and intentional design.

41) Amount of nitrogen in fossils:

Argument: The presence of nitrogen in fossils that are assumed to be millions of years old contradicts evolutionary timescales, since this element decays over time and should not be present in fossils this old.

Nitrogen is an element found in all living things. However, over time, this element decomposes and disappears. Fossils are the remains of plants or animals that lived a long time ago and turned into stone.

Some fossils are supposed to be millions of years old. This is a problem for evolutionary timescales, because if fossils were really that old, nitrogen should have completely decayed by now and should not be present. Therefore, finding nitrogen in ancient fossils suggests that they are not millions of years old, as claimed.

They try to refute with the following arguments:

1. "Not all nitrogen compounds break down at the same rate"

This is an argument that attempts to explain why nitrogen might remain in supposedly ancient fossils. However, while some forms of nitrogen are more stable, the fact is that the geological time to which evolution refers (millions of years) is long enough that even the most stable nitrogen compounds should have disappeared. Therefore, finding nitrogen in fossils is a strong indication that they are not millions of years old, but are much younger than the evolutionary paradigm suggests.

2. "Environmental conditions influence decomposition"

This argument suggests that nitrogen breakdown depends on factors such as temperature and humidity. However, the range of environmental conditions that would allow nitrogen to survive for millions of years is extremely limited. Most fossils are not found in such exceptional conditions, so it is highly unlikely that nitrogen has lasted that long. This reinforces the idea that the Earth and its fossils are younger than is claimed.

3. "There are other ways to measure the age of a fossil"

Although other dating methods such as radiocarbon and others are mentioned, it is important to note that many of these methods assume unproven premises, such as that the decay rates of certain elements have remained constant for millions of years. In addition, dating methods often give results that are inconsistent or contradictory to biblical chronology.

From a young Earth's perspective, these inconsistencies are best explained when the limitations and erroneous assumptions of radiometric dating methods are recognized.

Finally, from a creationist position, the presence of nitrogen in fossils is clear evidence that these are not as old as claimed. This fits with the idea of a young Earth, in which geological and biological processes have occurred over a period of thousands of years, not millions, let alone trillions, not even in rocks. The evolutionary assumptions that attempt to explain this phenomenon simply do not hold up in the face of observable evidence.

Sources that analyze the presence of nitrogen and other compounds in ancient fossils and the difficulty of maintaining such elements on timescales of millions of years:

Giem, P. A. (2001). Carbon-14 Content of Fossil Carbon. Origins, 51, 6-30. This article discusses the conservation of organic compounds in fossils and how their presence defies long time scales.

Snelling, A. A. (2008). Radiocarbon in "Ancient" Fossil Wood. Answers Research Journal, 1, 123-144. Snelling checks for the presence of compounds such as nitrogen and carbon in fossils, indicating that these fossils could be much younger.

Baumgardner, J. R. (2005). Carbon and Nitrogen Isotopic Composition of Fossils. RATE II, Institute for Creation Research. This work explores the implications of nitrogen found in fossils and suggests limitations in evolutionary theories about geological ages.

42) Oxygen in the atmosphere:

Argument: The level of oxygen in the atmosphere is kept in a delicate balance. If there were too much or too little oxygen, life on Earth could not exist as we know it. This balance is difficult to maintain for millions of years, which suggests that the Earth might not be that old.

Rather than this equilibrium remaining stable for long periods, it seems more logical that this system would have been functioning precisely for a much shorter time, which supports the idea of a young Earth.

A lot of oxygen spoils the broth. Not only is there no empirical evidence for "prebiotic broth" (the soup rich in organic compounds from which, according to evolutionists, life emerged), but there are other problems with this theory.

For example, it has been shown (read well, it has been scientifically proven, it can be demonstrated again) that any organic substance formed in the early days of the Earth would have been oxidized and quickly destroyed by the oxygen present in the atmosphere.

As a result, these organic compounds would not have had time to accumulate to form the "prebiotic broth."

Therefore, life would not have had the necessary conditions to emerge and develop in an oxygen-rich environment. This also calls into question the evolutionary theory of the origin of life, as oxygen would have prevented the building blocks for life from forming and persisting.

This contradiction reinforces the idea that the Earth is not as old as it is believed and that evolutionary models do not satisfactorily explain the origin of life.

Relevant sources on the stability of oxygen in the atmosphere and its impact on life and the origin of life:

Dent, J. N., & DiMichele, W. A. (1991). Oxygen and Evolution in Paleoatmospheres. Annual Review of Earth and Planetary Sciences, 19, 129-158. This article discusses how oxygen levels in the atmosphere have affected life at different periods and highlights the difficulty of maintaining a long-term balance.

Gonzalez, G., & Richards, J. W. (2004). The Privileged Planet: How Our Place in the Cosmos Is Designed for Discovery. It discusses fine-tuning oxygen levels in the atmosphere and how this supports the viability of life.

Summons, R. E., et al. (1999). Molecular evidence for Precambrian origins of chlorophyll and bacteriochlorophyll. Science, 283(5402), 366-368. This study refers to the challenges for the accumulation of organic compounds in the presence of oxygen, affecting the theory of "prebiotic broth".

43) The Ice Age at the Poles:

Plot: In places like Greenland and Antarctica, there are ice sheets that have been accumulating over time. Scientists say these sheets are millions of years old, but the amount of ice that accumulates each year is not enough for them to have formed in such a long time.

Think of ice layers as layers on a cake. Each year, a new layer of snow turns to ice, adding to the layers of previous years. If ice sheets really were millions of years old, there would be a lot more ice than we see today.

What we're seeing is that ice accumulation rates (the amount of ice that forms each year) don't jibe with the idea that ice sheets are so old.

In fact, it appears that the ice is much younger than has been claimed. This suggests that Earth is not as many millions of years old as is believed, and that ice sheets at the poles could have formed in a much shorter period of time.

Relevant sources on the accumulation of ice at the poles and its relationship with geological time:

Paterson, W. S. B. (1994). *The Physics of Glaciers*. This book provides in-depth insight into ice accumulation rates and the impact on the interpretation of ice sheet age.

Alley, R. B., et al. (2010). History of the Greenland Ice Sheet: Paleoclimate insights. *Science, 329(5993), 200-204*. This article reviews the ice record in Greenland and discusses how accumulation rates affect age models.

Oard, M. J. (2005). *Greenland Ice Cores: Implications for the Age of the Earth*. Institute for Creation Research. It critically examines ice sheet dating methods and proposes a more recent time frame.

44) The balance of CO_2 in the atmosphere.

Plot: Carbon dioxide (CO_2) is a gas essential to life on Earth. Plants need it to photosynthesize, and it also plays an important role in the planet's climate.

However, the current balance of CO_2 in the atmosphere is unstable in the long term.

This means that it cannot be maintained at the same levels for millions of years.

CO2 is absorbed by the oceans, plants, and other natural processes. If the Earth really were millions of years old, CO_2 would have been exhausted long ago or, conversely, it could have accumulated to such high levels that it would make life impossible. But we don't see that.

The level of CO_2 in the atmosphere has remained relatively stable in the times we can measure, which is more consistent with a young Earth than one that is millions of years old.

If the Earth were as old as evolutionists say, CO_2 would not be in the equilibrium we see today. This suggests that the Earth is much younger and that this balance has been maintained only for thousands of years, not millions. This argument highlights that the stability of CO_2 over millions of years is difficult to explain from an evolutionary perspective.

It is true that CO_2 is part of a natural cycle in which it moves between the atmosphere, oceans, plants and sedimentary rocks. However, this cycle has limits. The central argument of a young Earth is that if Earth had had millions of years for this cycle to occur uninterrupted, the accumulation of carbon in certain deposits (such as sedimentary rocks) would have led to a depletion of atmospheric CO_2, or to a level so high that it would affect life on Earth.

It is also true that plants and oceans play a key role in the absorption of CO_2. However, this system can also become saturated. But, if the Earth were millions of years old, the oceans would have absorbed far more CO_2 than they can sustain in solution, or soils and plants would have reduced CO_2 levels to dangerously low levels. The stability observed over thousands of years is much more consistent with a young Earth.

And..., although sedimentary rocks store carbon, and volcanic activity releases CO_2, this balance is delicate. If the Earth had been subjected to these cycles for millions of years, the balance would have been broken, either accumulating CO_2 uncontrollably or depleting it to levels that would not allow for life. Geological processes also have time limits that do not agree with millions of years of stability.

Analyses of ice cores have been interpreted by some scientists as evidence of natural fluctuations in CO_2 over hundreds of thousands of years. However, these dating methods also assume a long and stable time on Earth, which is debated. From a young Earth perspective, the CO2 fluctuations in these cores could correspond to recent weather events, such as Noah's Flood, which would have drastically affected CO2 levels.

Climate models are useful tools, but they are built on certain assumptions. They assume that climate conditions and carbon cycles have remained stable for millions of years. However, these models cannot accurately predict the long-term disruptions that might have occurred on a much younger Earth. In addition, models cannot directly prove the age of the Earth, they only simulate based on parameters that are already established.

The carbon cycle is a real process, but it is also finite and limited by several factors. If the Earth really were millions of years old, the current CO2 balance would not be as stable as we see it today. From a young Earth perspective, the observed stability of CO2 is more consistent with a much more recent story consistent with the biblical model of creation.

The following sources may be helpful:

Berner, R. A. (1994). Geocarb II: A Revised Model of Atmospheric CO_2 Over Phanerozoic Time. American Journal of Science, 294(1), 56-91. Berner discusses the long-term carbon cycle and its impact on CO_2 levels.

Vardiman, L. (2008). Ice Cores and the Age of the Earth. Institute for Creation Research. This article examines ice cores and how CO_2 levels can support a young Earth.

Holland, H. D. (2006). The Oxygenation of the Atmosphere and Oceans. Philosophical Transactions of the Royal Society B: Biological Sciences, 361(1470), 903-915. It analyzes how CO_2 and oxygen balance in the atmosphere, highlighting the fragility of this balance over long timescales.

45) Decreased Magnetic Field

Plot: Earth's magnetic field is weakening at a significant rate. If the Earth were billions of years old, the magnetic field should have already disappeared or reduced to very low levels. This suggests that the Earth is much younger than evolutionary theory proposes.

What is the Earth's magnetic field? Earth's magnetic field is a kind of protective "bubble" that surrounds the planet and protects us from harmful radiation from space, such as cosmic rays and solar storms. This field is generated by the movement of molten iron in the Earth's core.

What is happening with the magnetic field? Scientists have observed that the Earth's magnetic field is weakening. In fact, records show that it has been losing strength for the past few hundred years. If we follow the rate of weakening backwards in time, we conclude that, just a few thousand years ago, the magnetic field would have been much stronger than it is today.

The problem with a billion-year-old Earth: If Earth were billions of years old, this weakening of the magnetic field should have caused the field to have completely dissipated by now. Not only that, but life as we know it would have been much harder to sustain without a strong magnetic field to protect us from cosmic radiation.

Evidence for a young Earth: The fact that the magnetic field still exists and is strongly present suggests that it hasn't been there for millions or billions of years. From a young Earth perspective, the magnetic field has been weakening for only a few thousand years, which is consistent with a planet that is not that old.

The earth's magnetic field provides support for the idea of a young planet. A strong magnetic field is crucial for life as we know it.

It forms a protective covering around the planet that blocks the harmful cosmic radiation that continually bombards us.

Observations made of the earth's magnetic field over the past century and a half show that its intensity decreases perceptibly.

It is estimated that since 1829 the strength of the magnetic field has decreased by about 7%.

Calculations indicate that the half-life of the magnetic field is around 1,400 years, which means that it decreases to half its strength every 1,400 years.

Life would not be possible if it became too weak, and if the earth were so old the magnetic field would no longer exist or would have been so strong not long ago that it would have been impossible for life to develop.

Dr. Thomas Barnes, former dean of the Higher Institute of Creation and Professor Emeritus of Physics at the University of Texas at El Paso, calculates that at no time beyond 20,000 years would life as we know it have been possible.

The weakening of the magnetic field is inconsistent with the idea of a billion-year-old Earth.

Instead, the magnetic field appears to have existed for a much shorter period, which is compatible with the idea of a young Earth.

Barnes, T. G. (1973). Origin and Destiny of the Earth's Magnetic Field. Institute for Creation Research. Barnes argues that the magnetic field is decreasing at a rate incompatible with ancient Earth and suggests a shorter chronology.

Humphreys, D. R. (1984). The Earth's Magnetic Field is Young. Creation Research Society Quarterly, 21(3), 140-149. Humphreys examines the rapidly weakening magnetic field and proposes that the Earth is young.

Coe, R. S., Hongre, L., & Glatzmaier, G. A. (2000). An examination of simulated geomagnetic reversals. Philosophical Transactions of the Royal Society A, 358 (1768), 1141-1170. This study reviews the theory of the Earth's magnetic field, including fluctuations and weakening.

These sources explore the weathering of the magnetic field as evidence of a possible young Earth.

46) Weathering of Oceanic Crust

Argument: The oceanic crust, which is the thinnest layer of the Earth's crust, is being weathered due to tectonic activity.

If the Earth were billions of years old, the oceanic crust should have been completely eroded or worn away by now. However, the current thickness of the ocean's crust does not match such an old Earth, indicating that the planet is much younger.

What is oceanic crust? Oceanic crust is the rocky layer found under the oceans. It is thinner than continental crust and is constantly changing due to tectonic movements. The ocean floor is renewed when tectonic plates separate and magma rises to create new crust, a process that occurs at mid-ocean ridges.

What is happening to the thickness of the crust? Over time, oceanic crust is constantly renewing, but it is also subject to wear and tear. This wear and tear occur when an oceanic plate sinks under another plate (in a process called subduction), eventually reducing its thickness. Despite this process, the current thickness of the oceanic crust is still significant.

The problem with a billion-year-old Earth: If Earth really were billions of years old, this cycle of renewal and wear and tear would have caused the oceanic crust to thin much more than we observe today.

In other words, if oceanic crust has been forming and destroying for so long, it should be much thinner or even gone in some areas.

Evidence for a young Earth: The fact that the oceanic crust still retains considerable thickness suggests that this weathering process has not been occurring for billions of years.

From a young Earth perspective, the cycle of formation and weathering of oceanic crust has been going on for only thousands of years, which would explain why its thickness is still considerable.

The current thickness of the oceanic crust is incompatible with a billion-year-old Earth, as tectonic weathering should have drastically reduced its thickness. This suggests that the Earth is much younger and that the process of formation and weathering of the oceanic crust has been occurring in a much shorter period of time.

Some sources that address the renewal and weathering of oceanic crust, highlighting geological processes and the length of the tectonic cycle:

Staudigel, H., & Hart, S. R. (1983). Alteration of oceanic crust: Processes and timing. Earth and Planetary Science Letters, 58(1), 255-277. It analyzes the wear and renewal of the oceanic crust in relation to geological time.

Müller, R. D., Roest, W. R., & Royer, J.-Y. (1997). Digital isochrons of the world's ocean floor. Journal of Geophysical Research: Solid Earth, 102(B2), 3211-3214. Examina la antigüedad y renovación de la corteza oceánica.

Snelling, A. A. (2009). Earth's Catastrophic Past: Geology, Creation, & the Flood. Institute for Creation Research. This work addresses from a creationist perspective the renewal of the oceanic crust as evidence of a young Earth.

Part IV: Physical and Atmospheric Evidence

47) Comet Analysis:

Plot (and one of the ones I like the most, along with the distance from the moon): Comets are celestial bodies composed of ice and dust that, every time they approach the Sun, lose mass due to the sublimation of ice in their cores.

This process, known as "outgassing," generates the bright tails characteristic of comets.

However, since comets lose a significant amount of mass with each close pass to the Sun, they shouldn't last more than 10,000 years before completely disintegrating.

The continued presence of comets in the solar system poses a challenge to models that suggest our solar system is billions of years old.

If the solar system were this old, comets should have been exhausted by now.

To explain their persistence, "scientists" have proposed that the new comets come from a hypothetical reservoir known as the **Oort Cloud** and, to a lesser extent, from the **Kuiper Belt**.

However, this Oort Cloud has never been directly observed, which raises an important question: are they based on an "invention" to justify the longevity of the solar system? Is it valid to build a theory on something that does not exist in a proven way?

In this case, science seems to venture into the mystical, proposing a solution based on a hypothesis without direct evidence.

The continued existence of comets thus becomes intriguing evidence in favor of a shorter chronology for the solar system.

The existence of comets in the solar system supports the idea that the solar system is relatively young.

Sources of support:

Whipple, Fred L. (1950). "A Comet Model. I. The Acceleration of Comet Encke," The Astrophysical Journal, which analyzes the model of comets and their sublimation behavior.

Brandt, John C. & Chapman, Robert D. (1992). Introduction to Comets, Cambridge University Press, detailing the effects of sublimation and the life expectancy of comets.

48) Stellar Laughter.

Plot: White dwarf stars, which should have exhausted their fuel in tens of thousands of years, are still shining. This indicates that the universe could be younger than previously thought.

White dwarfs are the remnants of stars that have exhausted their nuclear fuel and collapsed under their own gravity.

According to astrophysical models, these stars should have cooled and stopped shining in a relatively short period in cosmic terms, in tens of thousands of years.

However, we still observe white dwarfs that emit light, suggesting that they haven't cooled enough to shut down.

Since white dwarfs should cool and shut down in relatively short times, their continued luminosity poses a problem for models of a universe thought to be billions of years old.

This has led some to question the age of the universe, suggesting that it could be much younger than traditionally estimated.

The persistence of white dwarf brightness is an indication that supports the idea of a young universe, in which these astronomical objects have not had enough time to cool completely.

A common counterargument from proponents of an ancient universe and models of chance is based on the prolonged cooling process of white dwarfs. According to these models, the cooling can extend for billions of years, since white dwarfs not only emit light due to their initial temperature, but also because the extreme collapse and density of these stars allows their radiation and cooling to last much longer than other models indicate. Indeed, the duration of cooling of white dwarfs is subject to specific conditions of their composition and density, which can cause some dwarfs to shine for much longer without contradicting the traditional chronology of the universe.

That white dwarfs can shine for billions of years faces a few flaws:

1. **Theoretical Models**: They depend on simulations that require many assumptions about the composition and behavior of white dwarfs over enormous periods.

   There is not always consensus on the accuracy of these assumptions.

2. **Limited Direct Observation**: Unlike other cosmic phenomena, observational evidence for extremely old white dwarfs is scarce, leaving a large part of the theoretical model without empirical confirmation.

3. **Uncertain composition**: The cooling rate depends on the exact composition and structure of each white dwarf, and factors such as core crystallization are not fully understood. These factors can cause estimates to vary considerably, weakening the idea that brightness can be sustained for billions of years uniformly in all cases.

This phenomenon has led some scientists and astrophysicists to explore alternative models to better understand cosmic chronology

In short, although the counterargument is well-founded, it is based on theoretical conditions that are not fully verified, which means that the debate about the age of these stars remains open.

Sources:

Kowalski, P.M., & Saumon, D. (2006). Physics of white dwarf atmospheres. This study looks at the cooling of white dwarfs.

D'Antona, F., & Mazzitelli, I. (1990). Cooling sequences of white dwarfs.

49) Geothermal Balance and Earth Cooling.

Plot: According to "scientific" theories, the Earth has been losing heat since its formation.

If the planet were billions of years old, as traditional evolutionary and geological models suggest, the Earth's core should have cooled completely.

However, the core is still extremely hot and active, resulting in phenomena such as volcanic activity and tectonic movement.

The persistence of a hot core, along with the geothermal energy that still emanates from the planet's interior, raises questions about the billion-year chronology proposed by evolutionists.

This internal heat on Earth might be more consistent with a much shorter timescale, which supports the idea of a young Earth.

In addition, the fact that internal heat still drives important geological processes, such as earthquakes and volcanic eruptions, bolsters the argument that the planet hasn't had enough time to cool completely, which would be more consistent with an Earth of a few thousand years rather than billions.

Scientific counterargument: Proponents of the ancient Earth model argue that the Earth's core keeps its temperature elevated due to radiogenic *heat*, generated by the decay of radioactive elements such as uranium, thorium, and potassium in the mantle.

Radiogenic **heat** is the heat produced by the **radioactive decay** of unstable elements within the Earth's mantle and core, such as uranium, thorium, and potassium-40. As these elements naturally break down, they release energy in the form of heat, which contributes to the Earth's internal temperature and helps keep geological processes active, such as volcanism and plate tectonic movement.

In addition, they argue that Earth's cooling is partially counteracted by convection processes in the mantle, which help keep heat inside the planet, thus contributing to observable geothermal activity. This waste heat could be explained within the proposed times of billions of years according to evolutionary theory.

Possible failure of the counterargument: A questionable point within the counterargument is the rate of decay of the radioactive elements and whether it is really able to sustain current heat levels for that long.

In other words, the question lies in whether the radioactive elements in the Earth's core and mantle, when they decompose and release heat, could really maintain sufficient internal temperature after billions of years.

Elements such as uranium and thorium, when they decay, generate heat, but this decay occurs at a rate that decreases over time.

Therefore, if the Earth were as old as evolutionary theory suggests, this heat generated by decay should have been greatly reduced, making it difficult to maintain the tectonic and volcanic activity observed today.

Also, the model of thermal cooling on a scale of billions of years would, in theory, lead to a significant reduction in tectonic and volcanic activity, something that is not consistent with the present geological activity.

To some researchers, these factors suggest a much more recent timeline for Earth.

To support the evidence on **geothermal equilibrium and radiogenic heat** in relation to the potential youth of the Earth, studies in geophysics and geology that address terrestrial waste heat, and the role of radioactive decay can be considered:

Vardiman, Larry - In his analysis *"The Age of the Earth's Atmosphere,"* Vardiman discusses how heat flow and dissipation of radioactive elements are inconsistent with an Earth billions of years old. The author argues that the rate of heat loss should have led to significant cooling millions of years ago, which does not seem to agree with the model of such an ancient Earth.

Snelling, Andrew A. - In his work *"Earth's Catastrophic Past: Geology, Creation & the Flood,"* Snelling reviews how geothermal activity (including that of volcanoes and earthquakes) indicates a more recent release of heat, pointing out that the internal heat would still be too high for an Earth of such antiquity.

Baumgardner, John - With Baumgardner's "Plates in Motion" model, he explores how subduction and tectonic movements cannot be easily sustained in the standard chronology of millions of years due to the theoretical cooling of an ancient crust and mantle.

50) "Fossilized Trees in Multiple Strata"

Argument: Fossil trees have been found in various parts of the world that are positioned vertically, traversing multiple layers of sediments or geological strata.

Fossil trees that traverse various geological strata, known as **polystratus trees**, have been found in different locations around the world, and some of the most well-known sites include:

1. **Yellowstone National Park, United States**: In the Specimen Ridge Formation, there are examples of polystratus trees found passing through multiple layers of volcanic ash, which has been interpreted in different ways, both in contexts of rapid deposition processes and in long fossilization processes.

2. **Joggins Formation, Nova Scotia, Canada**: This famous site features fossil trees that lie upright along several layers of coal and sediment, in strata spanning hundreds of meters.

3. **Wairarapa, New Zealand**: Polystratus trees have also been found in layers of volcanic ash and sediment. Here, the fossils show a preservation in several layers of volcanic deposits that would have accumulated at successive intervals.

4. **Saint-Étienne Region, France**: There are also finds of upright trees in coal strata and sedimentary rocks in this area, some of which span several meters deep in the strata.

These observations of polystratus trees have been the subject of debate. Some scientists interpret these fossils as evidence of rapid sediment deposition events, while others suggest long periods of time in which trees could have been preserved in their upright position before fossilizing.

These fossil trees, known as "polystratums," present a challenge to conventional geological theory, which holds that these strata formed slowly over millions of years.

According to this theory, each stratum represents an extended period of time in Earth's history.

However, a single fossilized tree that runs through several of these layers could not have stood for millions of years without decomposing before being buried by more sediment.

The existence of these fossilized trees suggests that the strata must have formed quickly, rather than slowly, which is more consistent with a catastrophic event such as a global flood, where large amounts of sediment were deposited in a short period.

The fossil trees found support the idea that geological processes do not always occur gradually but can be the result of rapid and massive events that deposit large amounts of sediment in a short time, which is consistent with the theory of a global flood described in the Bible.

Here are some recommended resources:

Snelling, Andrew A. - Earth's Catastrophic Past: Geology, Creation, and the Flood. This book addresses numerous examples of polystratus trees and discusses how they might be the result of catastrophic processes, supporting the hypothesis of a global flood.

Morris, John D. - Articles from the Institute for Creation Research (ICR). The *ICR* has published several papers analyzing polystratus tree finds at sites such as *Specimen Ridge in Yellowstone and Joggins in Nova Scotia.* These articles discuss how polystrata fossils pose challenges to conventional geological models based on uniformitarianism, i.e., the assumption that current processes are representative of the remote past.

Geological literature around the Joggins Formation, Nova Scotia. For investigations from a conventional perspective, many studies detail the deposition of sediment layers and fossils in the Joggins Formation.

The journal Geology and other academic resources can offer insights into the stratigraphy and context of these fossils.

Loma Linda University articles on Yellowstone and its geology. The university has published analyses of polystratus trees detailing their formation and interpretations of their preservation in rapid events, which could point to a catastrophic origin.

These resources allow us to explore both the creationist view and the analyses of conventional geology regarding the formation of polystratus fossils and their relevance in the context of the age of the Earth and sediment deposition.

51) The Magnetism of Rocks.

Plot: Magnetic tests on rocks of different geological ages show patterns that do not agree with models that propose that the Earth is billions of years old.

Magnetic tests on rocks, also known as paleomagnetism studies, analyze the magnetic field recorded in rock minerals during their formation.

Magnetic minerals, such as magnetite, record the orientation and strength of the Earth's magnetic field at the time of its crystallization. There are several tests and studies that are performed to investigate magnetic patterns:

1. **Reversal of the Magnetic Poles**: The Earth has experienced numerous magnetic field reversals, where the magnetic north and south poles change position. These reversals are recorded in rocks, especially those formed on mid-ocean ridges, where magma solidifies rapidly. As they cool, the magnetic minerals align with the moment's magnetic field, creating a pattern of alternating magnetic "bands" along the ocean floor.

   These patterns suggest rapid, recurring changes rather than a stable magnetic field for millions of years.

2. **Magnetic Field Strength and Orientation**: The magnetic field strength in rocks from different epochs does not always match what long-duration models predict.

   According to some models, the magnetic field should have dissipated considerably if the Earth were billions of years old.

A prominent author in this line is the physicist **Dr. Thomas G. Barnes**, who was a pioneer in the study of the decay of the Earth's magnetic field. In his work "Origin and Destiny of the Earth's Magnetic Field" (1973), Barnes argued that the Earth's magnetic field has a limited half-life and that its intensity has decreased significantly in recent times. According to Barnes, if this decay had been going on for billions of years, the magnetic field would have completely dissipated by now.

Researchers such as **Dr. Russell Humphreys**, associated with the Institute for Creation Research (ICR), have continued this line of thinking. Humphreys proposes that current data on the decrease in the magnetic field supports the hypothesis of a much younger Earth. Their models suggest that the current magnetic field is only a few thousand years old, in line with biblical chronology.

Publications in the *Creation Research Society Quarterly* and studies by organizations such as **Answers in Genesis** have also looked at the issue, questioning long timescales and arguing that decay models and the "residual intensity" of magnetism in ancient rocks align with a lesser-aged Earth.

These models have been debated in journals and conferences both within conventional geology and in critical geology forums, and references to their studies can be found in paleomagnetism papers in journals such as the *Geophysical Journal International*.

Studies on residual intensity in ancient minerals show a stronger field, which could be consistent with more recent geological history.

Studies showing higher residual intensity in ancient minerals have been conducted by scientists such as **Dr. Robert Coe** and **Michael Prevot**, who worked on paleomagnetism to understand the variation in the strength of the Earth's magnetic field over different geological epochs.

Another researcher, **Dr. John Baumgardner**, has also examined paleomagnetism in ancient volcanic rocks and his studies are in publications from the *Creation Research Society*

Quarterly, where he interprets the evidence as a possible indicator of a more recent geological history of the Earth.

In addition, research from the *Institute for Creation Research* (ICR), particularly studies led by **Dr. Russell Humphreys**, has reviewed the residual magnetic intensity in crystals of minerals such as zircon, stating that these data are consistent with a chronology of Earth of thousands of years, rather than millions.

3. **Magnetization Studies in Volcanic and Sedimentary Strata**: These studies analyze the remaining natural magnetization in volcanic layers or in ancient sediments containing ferromagnetic minerals.

   The variations found in the magnetization of rocks from different epochs are recorded and compared with the current pattern.

   Some research indicates that the "magnetic memory" of these rocks suggests more frequent and faster magnetic field fluctuations, which has raised questions about the duration of these processes and the age of the magnetic record.

4. **Magnetic Field Decay Models**: Another approach is to measure the decay of the magnetic field relative to geologic time.

   Some studies propose that, due to the wear and tear of the magnetic field, its current strength cannot be maintained stable for billions of years, since the energy of the core that supports the field would be exhausted in much less time.

These tests and methods have led to debates about the actual duration of the magnetic field and its relationship to the age of the Earth. For more details, paleomagnetism papers published in scientific journals such as *Geophysical Journal International* and *Journal of Geophysical Research* explore these phenomena and present data on the magnetic intensity and orientation of rocks.

Rocks, especially those containing minerals such as iron, register the Earth's magnetic field the moment they solidify.

This record, known as "magnetic remanence," has been studied to understand the history of Earth's magnetic field.

However, studies show that Earth's magnetic field has undergone significant fluctuations, suggesting that it has weakened at a much faster rate than expected if Earth were billions of years old.

In fact, if the magnetic field had existed for that long, it would have dissipated completely by now, leaving the Earth unprotected from solar radiation.

The fact that Earth's magnetic field has not completely dissipated can be interpreted as a case of **fine-tuning**, as its existence and characteristics seem specifically suited to sustain life on the planet.

Earth's magnetic field acts as a "shield" against solar radiation and solar winds, protecting both the atmosphere and the planet's surface from lethal exposure to these high-energy particles.

In the context of fine-tuning, this phenomenon suggests that conditions are not only in equilibrium but are specifically "tuned" to allow life.

The stability of the magnetic field, despite a process of continuous weakening, appears to be calibrated to remain at a level that remains suitable for life, although the exact mechanism of that stability is still under investigation.

Some scientists who analyze fine-tuning on Earth and in the universe have interpreted this phenomenon as an indication that natural, physical, and environmental factors seem to be configured in a precise and unique combination, favoring life on our planet. This contrasts with the expectation of a more chaotic and unstable system in a universe that would arise solely by chance.

Studies such as those of **Dr. Russell Humphreys** and reports from the *Institute for Creation Research (ICR)* discuss this "tuning" of the magnetic field in the context of a young Earth. Also, general research on fine-tuning, such as that published in cosmology and physics journals, addresses how the stability of these fields, in synchrony with other cosmic factors, raises interesting questions about the origin and chronology of our planet.

The *Institute for Creation Research* (ICR) has a team of scientists and researchers with specialties in biology, geology, paleontology, astrophysics, and other areas relevant to creation science. Although the exact numbers of scientists may vary, the ICR typically has an active base of about 10 to 15 full-time researchers and many contributing collaborators and authors. In addition, similar organizations, such as *Answers in Genesis* (AiG), which develops creationist research and educational projects, also have a team of experts and collaborators, including scientists with PhDs in various disciplines.

Most of these organizations are not recognized as part of the mainstream "scientific community" because their approach includes interpretations based on a literal biblical perspective.

In addition, rocks have been found to indicate rapid and abrupt reversals of the magnetic field, which contradicts models of geological evolution that propose gradual changes over millions of years.

These rapid fluctuations and steady decline in the magnetic field are most consistent with a young Earth, which is only a few thousand years old.

This challenges the idea of an ancient Earth and supports the notion that geological processes may have occurred more rapidly and recently than previously thought.

52) "Meteorite Shortage in Ancient Layers"

If Earth were really old, we'd expect to find a large number of meteorites embedded in ancient geological layers, but these are surprisingly scarce.

This suggests a more recent timeline for sediment accumulation and the formation of such layers.

The use of meteorites to calculate the age of the Earth *poses serious problems*, as these bodies did not originate directly from our planet.

The meteorites have traveled through space and could have been exposed to different levels of radiation and other cosmic conditions, affecting their composition and making their usefulness in accurately establishing Earth's age questionable.

This fact casts doubt on the results obtained by dating meteorites, which propose very old ages for the Earth.

In addition, meteorites continuously fall on the Earth's surface, and interestingly, they are only found in the upper layers of the sediment. If the deepest geological layers had formed millions of years ago, as evolutionists suggest, meteorites should be expected to be found in all layers of sediment. However, the absence of meteorites in the oldest strata indicates that these sediments are much younger than claimed. This reinforces the recent chronology of sediment formation and supports the idea of a young Earth.

It is interesting to highlight and even insist again since this is evidence that I consider solid, on several points related to the dating of the Earth and the deposition of meteorites in geological time.

1. **Problems in Meteorite Dating**: Meteorites, coming from space, have been exposed to different conditions than on Earth. These include varying levels of cosmic radiation and possible chemical alterations during their long journeys through space. These conditions can modify its isotopic composition, which makes it less accurate to extrapolate its age to date the Earth. Discrepancies in the radioactive decay of isotopes in meteorites have been questioned in dating studies and it is suggested that they do not necessarily reflect terrestrial chronology. This argument has been addressed in research by the Institute for Creation Research (ICR), such as *Russell Humphreys*' work on helium on Earth and the accuracy of radiometric methods.

2. **Surface Distribution of Meteorites**: Most meteorites are found in recent geological layers, especially in areas such as deserts and ice sheets in Antarctica, which accumulate meteorites on the surface.

In deep geological strata, meteorites are virtually non-existent, which begs the question: why is there no evidence of layered meteorites that supposedly formed millions of years ago?

This could be interpreted as an indication that the deep geological layers are not as old as suggested, or that the formation process of these layers was much more recent or faster than conventional models indicate.

3. **Implications for Young Earth Geology**: If a significant absence of meteorites is observed in the older layers, it can be argued that they did not go through long periods of sedimentation, but formed in a brief and catastrophic period, consistent with models that contemplate high-depositional events, such as Noah's Flood in creationist geology.

Researchers in this field, such as *Andrew* Snelling in *Earth's Catastrophic Past*, address how these events may have quickly created layers of sediment, leaving few or no meteorites deep in the depths due to the rapid deposition of terrestrial and aquatic materials.

4. **Counterarguments to Conventional Models**: Proponents of ancient Earth argue that the lack of meteorites in deeper layers could be due to degradation, oxidation, or subduction of materials in tectonically active areas.

However, these processes do not occur uniformly across the planet, and there should still be evidence of meteorites in more geologically stable areas.

Suggested sources:

Humphreys, D. Russell. "Helium Evidence for a Young Earth." *Institute for Creation Research*, exploring the influence of spatial conditions on isotope decay.

Snelling, Andrew A. Earth's Catastrophic Past: Geology, Creation & the Flood, which delves into geological models of rapid deposition in strata.

53) "Inconsistencies in Isotope Dating in Volcanic Rocks"

Argument: Radioisotope dating methods, which are commonly used to calculate the age of meteorites and volcanic rocks, have serious inconsistencies.

An example is the Allende meteorite, where multiple tests have shown contradictory results, which has led to the discarding of some samples for "contamination". This type of error generates a considerable margin of doubt and calls into question the reliability of these methods to accurately date the age of materials.

Isotope dating depends on several key assumptions, such as consistency in decay rate and sample purity, which may vary or not be met. These assumptions are difficult to verify, especially in the case of materials that are very old or altered by environmental factors.

Studies on meteorites such as Allende have shown that age results can vary widely depending on the **isotopes and measurement methods** used. This suggests that, rather than being absolute, these ages may be subject to interpretation according to starting models, which is central to the debate about the actual age of the Earth.

An even more obvious case is the dating of volcanic rocks from recent eruptions, such as those in Hawaii and Salt Lake. Although it is known with certainty that some eruptions occurred in recent historical times (even less than 200 years ago), rock samples from these eruptions have yielded ages of hundreds of thousands and even millions of years when methods such as potassium-argon (K-Ar) are applied.

In Salt Lake (certain studies on volcanic rocks from Utah and Nevada), similar cases have been found, where recent volcanic rocks exhibit age discrepancies in millions of years in isotopic dating results. The most commonly cited reason for these differences is attributed to the presence of **"excess argon"** or a **"contamination"** of the samples. This extra argon could come from a variety of geological sources and cannot be distinguished from argon produced by radioactive decay, which affects the accuracy of the measurement.

We know that these rocks formed just a few centuries ago, but radioisotope dating has yielded ages ranging from hundreds of thousands to millions of years. These huge discrepancies suggest that isotope dating methods are not as reliable as believed, and that their results may be highly inaccurate.

This raises questions about their use to calculate the age of the Earth, supporting the idea that its age could be much more recent than these methods indicate.

Faure, G., & Mensing, T. M. (2005). "Isotopes: Principles and Applications". Although this is a standard text in geochronology, it also acknowledges that dating results may vary due to factors such as contamination or alteration of samples, which are more difficult to control in some types of formations.

54) Polonium Halos in Rocks

Radioisotope dating methods, commonly used to calculate the age of meteorites and volcanic rocks, have serious inconsistencies. A clear example is the case of the Allende meteorite, where multiple tests have shown contradictory results, which has led to the discarding of some samples for "contamination". These errors create a considerable margin of doubt and call into question the reliability of these methods for accurately calculating the age of materials.

Evidence of "polonium halos" in granite rocks has been studied extensively by researchers such as Dr. Robert Gentry, who argued that these halos could not have formed under slow-cooling conditions.

According to their studies, polonium-218, which has a half-life of only 3 minutes, would have left visible patterns only if the rock had solidified almost instantaneously, which is inconsistent with the idea that granite formed slowly millions of years ago.

Gentry detailed his findings in various publications, highlighting that these halos remain an "unsolved mystery" within the scientific community, since no "parent element" has been found that explains the origin of polonium in these contexts. especially by the *Institute for Creation Research* (ICR).

This organization argues that the preservation of these halos is evidence of rapid geological cooling, which would fit into a model of recent creation.

However, the interpretation of these data has been controversial, and the mainstream scientific community considers it necessary to apply additional models to explore other possible explanations.

An even more obvious case is the dating of volcanic rocks from recent eruptions, such as those in Hawaii and Salt Lake. We know for sure that these rocks formed just a few centuries ago, however, radioisotope dating tests have yielded ages ranging from hundreds of thousands to millions of years.

These large discrepancies suggest that isotope dating methods are not as reliable as is often believed.

This raises important questions about their use to determine the age of the Earth, reinforcing the possibility that its age is much more recent than these methods indicate.

Complementary evidence is the "polonium-218 halos" found in granite rocks.

These halos are small circular patterns formed by the decay of this radioactive isotope, which has an extremely short half-life of only 3 minutes.

For these halos to be preserved, the rock must have solidified almost instantaneously.

This contradicts the idea that granite rocks formed slowly over millions of years, as evolutionary models suggest.

What's more intriguing is that no traces of a "parent element" that originated Polonium-218 are found in these cases, which is highly unusual and poses an unsolved mystery for scientists who sustain a billion-year-old Earth.

This phenomenon should not exist in ancient rocks, as polonium should have decayed long before the rock solidified.

Thus, the presence of these polonium halos in granite offers strong evidence that the Earth cooled rapidly and in a much shorter period than evolutionary models suggest. This reinforces the idea of the recent and rapid creation of the Earth.

Part V: Biological, Geological and Population Evidence.

In this section, we present more evidence that defies evolutionary timescales and suggests that humanity and life on Earth are much more recent than is unofficially claimed.

Through historical, genetic, fossil, and demographic records, evidence is explored that supports a chronology consistent with the biblical view of recent creation, calling into question the long periods of time proposed by evolutionary models.

55) Oldest historical records:

Plot: The oldest cultures have no records prior to 6,000 years, which matches biblical chronology and challenges the idea of a humanity that has existed for hundreds of thousands of years.

Older human civilizations, such as Sumer, Egypt, and the Indus Valley civilization began to develop writing systems and social organization approximately 5,000 to 6,000 years ago. Prior to that period, there is no archaeological evidence of writing systems or historical records documenting human life.

As for cities that might have been submerged, such as the underwater ruins of Yonaguni in Japan and some sites in India, which some suggest would date back as much as 10,000 years, they have been interpreted in both historical and mythological contexts.

This raises an important question: if humanity has existed for hundreds of thousands of years, why don't we find traces of advanced cultures or social structures that predate this period?

Conventional archaeology holds that anatomically modern humans (Homo sapiens) have been around for about 200,000 years, based on fossil remains and stone tools.

However, the lack of evidence of organized civilizations or historical records dating back beyond 6,000 years creates an inexplicable gap in this narrative. And between 6,000 years and 200,000 years there is a huge difference.

If humanity had existed for so long, one would expect to find tangible evidence of organized societies or older technological advances.

This archaeological gap agrees with biblical chronology, which places the origin of humanity and the development of the first civilizations around 6,000 years ago. Biblical chronology not only presents a coherent explanation for the absence of earlier historical records, but it also challenges the evolutionary view of human antiquity.

While some argue that many records could have been lost to the passage of time or the destruction of perishable materials, humans also used stone, pottery, and metals, which are more durable.

If older advanced civilizations exist, it would be expected to find more preserved physical evidence, but to date no such evidence has been found.

In addition, hominin fossils such as "Lucy" are interpreted within an ancient Earth framework using methods such as carbon-14 and radiometric dating.

These methods, from a creationist perspective, are subject to assumptions that can be questioned.

In addition to *"Lucy"* (Australopithecus afarensis), other hominid fossils that are considered ancient and are used in the evolutionary framework to explain human ancestry include:

1. **Ardi** (*Ardipithecus ramidus*): A fairly complete skeleton about 4.4 million years old found in Ethiopia. It is considered a close ancestor of the lineage that would have given rise to humans and stands out for its characteristics of bipedalism. *Ardi* is composed of fragmented bones from multiple parts of the skeleton, and its reconstruction has been complex. The structure of the skull and pelvis indicates that it is not a direct ancestor of humans but rather a primate.

2. **Java Man** (*Homo erectus*): Discovered in Indonesia, it dates to around 1.8 million years ago. This fossil has been key in the theory of the dispersal of the first hominids out of Africa. The

fossil remains found mainly include a partial skull, a molar, and a femur. It has been argued that the skull, due to its intermediate characteristics, could belong to a primate species closer to apes.

3. **Peking Man** (*Homo erectus pekinensis*): Found in China and also belonging to *Homo erectus*, it is estimated to be between 500,000 and 800,000 years old. The remains found include several skulls and teeth, as well as some facial bones and fragments of limbs. Regarding the availability for independent analysis, several of these fossils were reported as lost during World War II, so only casts and photographs of some original samples exist.

4. **Taung child** (*Australopithecus africanus*): A skull of a young man found in South Africa, approximately 2.8 million years old, which has been interpreted as evidence of brain evolution in hominids.

5. **Homo habilis**: This fossil, found in East Africa, is between 1.4 and 2.4 million years old. It has been associated with the use of primitive tools. Access to these remains has been limited, as the original fossils are mostly housed in museums and research labs and are not always available for analysis by all researchers, including those who support a creationist perspective.

6. **Neanderthals** (*Homo neanderthalensis*): Although more recent (between 40,000 and 400,000 years ago), Neanderthal fossils in Europe and Asia are important in the discussion of human ancestry. Studies suggest that they shared some cognitive abilities with present-day humans,

There has been some debate about the availability of fossil samples for outside studies, including researchers with creationist views. However, access has been possible in some institutions and publications, allowing experts in paleoanthropology to analyze the characteristics of these remains under various scientific approaches.

Each of these fossils is interpreted in the context of long periods of time and the theory of gradual evolution. However, from the perspective of a young Earth and with alternative dating methods, some suggest that the anatomical and geological features of these fossils could be compatible with much more recent chronologies.

Similarly, modern genetics, which suggest a common ancestor for all humans, is compatible with the idea that we are descended from a small recent group, such as Adam and Eve or the survivors of the Flood.

In short, the lack of tangible evidence of older civilizations and the sudden emergence of organized cultures about 6,000 years ago reinforce biblical chronology and cast doubt on the evolutionary narrative of a much older humanity.

The work "The Genesis Flood" by Henry M. Morris and John C. Whitcomb, which argues that many geological and archaeological phenomena are compatible with a global flood.

56) Accelerated evolution in fossils:

Many fossils found show a striking resemblance to present-day species and no evidence of gradual evolution over millions of years.

This fact questions the long evolutionary times and favors the idea of a recent creation, in which species were designed to remain stable over time.

"Living fossils," species that have remained almost unchanged, offer a clear example of biological stability.

Evolutionary theory suggests that species gradually evolve through intermediate forms, however, the absence of these transitional forms in the fossil record raises serious questions about the proposed evolutionary timings. Darwin himself pointed out this deficiency in his writings: "No change from one species to another is recorded... we cannot prove that a single species has been changed."

—Charles Darwin, *My Life and My Letters*, 1905.

In turn, in *On the Origin of Species*, Darwin recognized the lack of intermediate evidence as a major problem: "The number of intermediate varieties that existed in the world in ancient times should have been really enormous. Why, then, is not every geological formation present and every stratum full of intermediate links? Geology certainly does not reveal any organic change so finely gradual, and this is perhaps the most obvious and serious objection that can be raised against this theory."

—Charles Darwin, *The Origin of Species*.

Similarly, paleontologist Richard Leakey, a recognized figure in the research of human evolution, acknowledged the lack of transitional fossils in his specialty: "If you pressed me about the descent of man, I would have to say unequivocally that what we possess is only a big question mark. To date, nothing has been found that can be reliably supported as a transitional species towards man... If I were pressed further, I would have to declare that there is more evidence to suggest an abrupt arrival of man than a process of gradual evolution."

—Richard Leakey, PBS documentary, 1990.

The stability observed in living fossils and the lack of transitional fossils raise questions about the plausibility of gradual change over millions of years. Darwin and Leakey's statements reflect an unresolved observation: the absence of clear evidence of a continuous and gradual evolutionary process. These factors, not finding a conclusive answer, strengthen the position that life could have originated recently and deliberately, with designs that have remained stable over time.

Morris, H. M., & Parker, G. E. (1982). What is Creation Science? Master Books. This book analyzes the stability of species in the fossil record and the difficulty of finding transitional forms, arguing for creation and questioning long evolutionary times.

Gish, D. T. (1995). Evolution: The Fossils Still Say No!: Gish explores cases of "living fossils" and the lack of transitional fossils in the geologic record, highlighting species that show no significant change over the purported millions of years.

Wise, K. (2002). Faith, Form, and Time: What the Bible Teaches and Science Confirms about Creation and the Age of the Earth. Broadman & Holman.: Este autor discute cómo la evidencia fósil puede ser más coherente con un modelo de creación que con los cambios graduales propuestos por la teoría evolutiva. Analiza especies actuales y fósiles casi idénticos en su forma a lo largo del tiempo.

Eldredge, N., & Tattersall, I. (1982). The Myths of Human Evolution. Columbia University Press.: Aunque Eldredge y Tattersall son evolucionistas, este libro reconoce la falta de "eslabones perdidos" en el registro fósil humano y menciona cómo la evidencia de transición es a menudo difícil de interpretar y escasa.

These sources address the problem of gradual evolution and the lack of intermediate species in the fossil record from various perspectives, supporting the stability observed in species as possible evidence in favor of a recent creation.

57) Genetic material in ancient fossils:

The finding of DNA and proteins in fossils of dinosaurs and other species that are supposedly millions of years old suggests that these fossils are much more recent than claimed.

Science states that both DNA and proteins break down rapidly within thousands to tens of thousands of years due to processes such as oxidation and hydrolysis.

To support the claim that DNA and proteins break down in thousands or tens of thousands of years, key studies have shown the impact of natural processes such as oxidation and hydrolysis on the degradation of these organic materials:

- ✓ **Lindahl, T. (1993)** - Biochemist Tomas Lindahl is one of the pioneers in DNA degradation research. In his study, *Instability and Decay of the Primary Structure of DNA*, published in *Nature*, Lindahl showed how DNA undergoes irreversible changes due to processes such as oxidation, hydrolysis and deamination. Lindahl conducted experiments to observe how quickly DNA degrades under natural conditions, concluding that it can only persist intact for up to tens of thousands of years.
- ✓ **Hofreiter, M., et al. (2001)** - In the Ancient *DNA study*, published in *Nature Reviews Genetics*, Hofreiter and his team examined the survival of DNA under ideal conservation conditions. Their findings determined that decay occurs rapidly under environmental exposure, and although the survival of DNA in fossils can be extended under freezing conditions, under normal temperature conditions it completely degrades in less than 100,000 years.
- ✓ **Collins, M. J., et al. (2002)** - Matthew Collins' team in the study *The Survival of Organic Molecules in the Fossil Record*, published in *Philosophical Transactions of the Royal Society B*, explored how proteins, especially collagen, are degraded in fossils. This study employed dating techniques and fragmentation analysis to observe that proteins tend to lose

their structure and functionality over thousands of years, even in well-preserved fossils.

These studies have been instrumental in establishing time limits for the preservation of DNA and proteins, showing that such materials cannot last for millions of years, as some dinosaur fossils suggest, but that the persistence of these components is much more consistent with a recent time frame.

These processes are inevitable even under favorable conditions of preservation.

However, in recent years, traces of DNA, collagen, and other organic remains have been found in fossils of dinosaurs and other species estimated to be tens or even hundreds of millions of years old.

The widely accepted hypothesis in the scientific community suggests that dinosaurs went extinct approximately 66 million years ago due to a catastrophic event triggered by a meteorite fall, known as the Cretaceous-Paleogene (K-Pg) extinction event. This impact would have occurred in what is now the Chicxulub crater, in the Yucatan Peninsula, Mexico. The date of the impact does not exactly coincide with the disappearance of all dinosaurs, and the size of the crater does not fully explain the degree of extinction. The lack of a uniform "disappearing point" for all species calls into question whether the impact was the ultimate cause.

Proponents of biblical chronology (like me) are of the opinion that the extinction of the dinosaurs could be related to the account of the Flood. Noah's Flood would have been a global catastrophic event that altered the Earth's climate, terrain, and ecosystem as a whole, causing the disappearance of many species, including dinosaurs.

Continuing the argument, the findings pose a challenge to traditional evolutionary chronology, since, according to current knowledge, organic macromolecules should not survive that long. Some examples include:

> ➢ **Soft tissue and collagen discoveries** in fossils of Tyrannosaurus rex and other species, made by paleontologist Mary Schweitzer, who found blood vessels and collagen in dinosaur bones that were supposed to be 68 million years old.

> **Protein and DNA remains in fossils of mammoths and other more recent specimens** from tens of thousands of years ago, which have generated debates about the real limits of preservation of these materials.

The discovery of these intact organic materials in ancient fossils suggests that such fossils could be much more recent than claimed. This supports the idea of a young Earth and raises significant questions about traditional evolutionary timescales.

Schweitzer, M. H., et al. (2007). "Soft tissue vessels and cellular preservation in Tyrannosaurus rex." Science, 316(5822), 277-280.

Lindgren, J., et al. (2011). "Molecular preservation of the pigment melanin in fossil melanosomes." Nature Communications, 2, 417.

58) Fossil remains without evolutionary change:

This argument posits that certain fossils, often called "living fossils," show a remarkable similarity to their present-day counterparts, which seems to contradict the theory of gradual evolution. If the evolutionary process were continuous and changes accumulated over time, one would expect to see more marked differences between fossils and modern species.

However, examples such as the coelacanth (a fish considered extinct until its rediscovery in 1938), the horseshoe crab, or certain species of sharks, suggest genetic and morphological stability.

Evolutionary theory explains the existence of these "living fossils" through "evolutionary stasis," suggesting that, in certain stable environments, some organisms have not needed to change significantly.

However, the continued discovery of these fossils in varied contexts and periods has led some to consider that these examples of biological stability fit better into a recent chronology, where organisms were created with their current complexity and have not changed radically over time.

For supporters of a young Earth, this lack of visible evolution could be interpreted as confirmation that species were designed with a great capacity to adapt within specific limits, without the need for continuous transformations at the evolutionary level.

This stability is also used as evidence in favor of creation, in which the biodiversity and functionality of current species would have been established in a recent period, without long and uninterrupted processes of change; that is, during the process of creation described in Genesis.

59) Limited genetic diversity:

Argument: The relatively low genetic diversity in the human population indicates that humanity has not had millions of years to accumulate large genetic differences, but that we are probably descended from a small number of individuals in a recent period.

The low genetic diversity in the human population suggests a recent origin of humanity.

If humanity had existed for millions of years, genetic differences between individuals would be expected to be much more pronounced due to the long time available for mutation accumulation.

However, genetic evidence indicates that humans are likely descended from a small number of individuals in a recent period, which is more consistent with a recent event, such as a population bottleneck, and supports the idea of a shorter timeline for humanity.

The genetic differences between people are not as great as we would expect if humans had existed for millions of years. If humanity were this old, there would have been much more time for our genetic differences to get bigger.

But what we see is that people around the world are quite similar in genetic terms. This suggests that we are all descended from a small group of people relatively recently, which fits better with the idea that humanity is much younger than is claimed.

Relevant sources discussing evolutionary stability in organisms, also known as "stasis":

Eldredge, N., & Gould, S.J. (1972). "Punctuated Equilibria: An Alternative to Phyletic Gradualism." In Models in Paleobiology, he explains stasis in the fossil record as a phenomenon observed in several species, where many do not exhibit significant changes over long geological time scales.

Benton, M.J. (2001). "Finding the tree of life: matching phylogenetic trees to the fossil record through the 20th century." Proceedings of the Royal Society B: Biological Sciences, highlights cases such as the coelacanth and the horseshoe crab, which have shown remarkably prolonged morphological stability.

Stanley, S.M. (1981). The New Evolutionary Timetable: Fossils, Genes, and the Origin of Species, describes how certain organisms have undergone little or no morphological evolution over millions of years, an observation that raises questions about variability in evolutionary rates.

Gould, S.J. (2002). The Structure of Evolutionary Theory argues for evolutionary stasis and suggests that evolutionary change is not always gradual, highlighting the lack of change in "living fossil" species.

These sources offer both observational data and theoretical arguments about the persistence of certain organisms without significant changes, which allows us to question the model of gradual evolution in favor of hypotheses of stasis or genetic stability in certain species.

60) Distribution of ancient languages:

Argument: The world's major language families all seem to have developed in the last 4,000 to 5,000 years, which coincides with the dispersal of peoples described in the Bible and suggests that humanity has not had hundreds of thousands of years to diversify its languages.

This coincides with the biblical account of the dispersion of peoples, such as the Tower of Babel event. If humanity had existed for hundreds of thousands of years, one would expect languages to have shown much greater diversification.

However, the recent formation of these language families suggests that humanity has not had as much time to diversify its languages, which supports the idea of a shorter chronology for humanity.

The estimate that the most recent common ancestor (MRCA) of all modern humans lived approximately 200,000 years ago comes from genetic studies, specifically mitochondrial DNA and chromosome

And. However, it is important to clarify that this estimate refers to a genetic ancestor and does not imply that modern languages originated at that time.

The development of languages is a cultural process, not a genetic one, and depends on the history of human civilizations, the dispersion of peoples, and the contacts between them. Although genetic studies suggest an age of 200,000 years for the MRCA, linguistic and archaeological evidence indicates that the world's major language families developed only about 4,000 to 5,000 years ago, which is more in line with the account of dispersal described in the Bible and with a recent development of human languages.

Therefore, although genetic studies propose a longer timeline for the common origin of modern humans (an unlikely issue as we have seen elsewhere), this does not contradict the idea that the development of languages has occurred much more recently. Languages and culture are not governed by the same rules as genetics, and their evolution may have been more rapid and recent.

Ruhlen, M. (1994). On the Origin of Languages: Studies in Linguistic Taxonomy.

Nettle, D., & Romaine, S. (2000). Vanishing Voices: The Extinction of the World's Languages.

61) Lack of ancient human remains:

The lack of human remains dating back hundreds of thousands of years raises an important question as to the chronology of humanity.

According to evolutionary theory, the anatomically modern human being (Homo sapiens) has existed for approximately 200,000 years.

If true, one would expect a significant amount of human remains in geological layers corresponding to that long period, which is not observed in the current fossil record.

Archaeological excavations, which have revealed fossils of other species and human cultures several millennia old, have been insufficient to find human fossils in quantities and depths corresponding to hundreds of thousands of years.

This has led to questions about the coherence of evolutionary timescales, which suggest human presence for hundreds of thousands of years, since most of the human remains found date back to recent times.

A shorter chronology, such as that suggested by the model of a young Earth, better explains the scarcity of human remains in ancient geological layers, since, in this context, humanity would not have existed long enough to accumulate traces in large quantities in geological strata.

Tattersall, I., & Schwartz, J. H. (2008). The Human Fossil Record.

Klein, R. G. (2009). The Human Career: Human Biological and Cultural Origins.

62) Distribution of civilizations:

The emergence and rapid expansion of ancient civilizations, such as Sumer, Egypt, and the Indus Valley civilization, all of which emerged less than 6,000 years ago, supports a recent timeline for the development of human societies.

These civilizations developed and prospered rapidly, establishing complex social structures, writing systems, and technological advancements in a short period.

If humanity had existed for hundreds of thousands of years, one would expect advanced civilizations to emerge much sooner.

However, archaeology and historical records point to a relatively recent beginning of organized civilization, which supports a shorter chronology and questions long evolutionary timescales.

Durant, W. (1954). Our Oriental Heritage: The Story of Civilization.

Oppenheim, A. L. (1964). Ancient Mesopotamia: Portrait of a Dead Civilization.

Hawkes, J. (1973). The First Great Civilizations: Life in Mesopotamia, the Indus Valley, and Egypt.

63) "Potassium Isotopes and Volcanic Rocks"

Potassium isotopes, particularly potassium-40 (K-40), are commonly used to date volcanic rocks and are considered to decay in a predictable process, transforming into argon-40 (Ar-40).

The decay of K-40 has a half-life that, in theory, allows scientists to estimate the age of rocks on timescales of millions of years.

However, studies have shown that geological conditions can affect the stability of this isotope and alter the results of dating tests.

Challenges in Potassium-Argon (K-Ar) Dating: The K-Ar method has come under fire for the unusually high results it produces when analyzing recent volcanic rocks.

Studies of known eruptions, such as Mount St. Helens (1980) and some in Hawaii, have shown ages ranging from thousands to millions of years, although these rocks are only a few centuries old.

This discrepancy raises questions about the reliability of K-40 dating and its ability to accurately reflect the age of rocks.

The reason for these problems may be related to argon trapping and the variability in the rate of K-40 decomposition under specific pressure and temperature conditions.

- **Mount St. Helens:** Rock samples produced during the 1980 eruption, dated by the K-Ar method, resulted in ages of up to 350,000 years. This type of error means that, instead of reflecting the true age of the rock, K-Ar dating is susceptible to incorporating amounts of trapped argon, creating an artificially high reading.
- **Eruptions in Hawaii:** Studies of rocks from recent eruptions in Hawaii, specifically at Kilauea volcano, have yielded ages of up to 3 million years in some samples. These dates do not match the documented history of these eruptions, indicating that argon buildup in the rocks affected the readings and

suggesting that K-40 may not be a reliable indicator in all circumstances.

Evidence for a Recent Crop: These inconsistencies in potassium-argon dating pose serious problems for the evolutionary model of an ancient Earth, where radioactive isotopes supposedly allow events to be dated on scales of millions of years.

The presence of extra argon and the variability in K-40's decay suggest that current methods may overestimate the actual age of volcanic rocks.

From a young Earth perspective, these results may indicate that the isotopic methods are not as stable or as accurate as presented, which reinforces the idea of a more recent terrestrial chronology.

References:

Austin, Steven A. *Excess Argon witl Concentrates from the New Dacite Lava Dome at Mount St. Helens Volcano*. Creation Research Society Quarterly, 1996.

Snelling, Andrew A. *Radioisotopes and the AEarth: A Young-Earth Creationist Research Initiative*. Institute for Creation Research & Creation Research Society, 2005.

Dalrymple, G. Brent, *Potassium-Argon Dating: Principiques, and Applications to Geochronology*, W.H. Freeman, 1975.

Hayatsu, Akira. *Argon-Retention and the Problem of Dating Youc Rocks*. Geochimica et Cosmochimica Acta, 1979.

This argument shows how inconsistencies in potassium-argon dating can call into question the chronology of an ancient Earth, suggesting a dating alternative compatible with a more recent geological history.

64) Cambrian Explosion Theory:

Plot: The fossil record shows that a large number of complex species suddenly appeared during the Cambrian period, without the expected evolutionary precursors.

This event, known as the "Cambrian Explosion," poses serious challenges to theories of gradual evolution and suggests that life on Earth could have been created in a short period of time.

This is a classic in debates about the origin of life and the theory of evolution. It is based on the observation that during the Cambrian period there was a rapid diversification of life forms, many of them complex, with no clear evidence of intermediate life forms connecting them with simpler organisms.

The "Cambrian Explosion" is a well-documented event in the fossil record, where a large number of complex species emerged suddenly, with no clear transitional forms showing gradual evolution.

According to evolutionary theory, we should observe evolutionary precursors of these complex species, but in the fossil strata prior to the Cambrian, these precursors are practically non-existent.

This event poses serious challenges to theories of gradual evolution, as the rapid emergence of such diverse and complex life forms does not fit the model of slow and cumulative change that evolution proposes.

Instead, this phenomenon is more consistent with the idea that life could have been created in a short period of time, supporting the possibility of a recent creation and not a long, slow process of evolution.

Sources discussing the "Cambrian Explosion" and its implications for evolutionary theory include:

Meyer, S. C. Darwin's Doubt: The Explosive Origin of Animal Life and the Case for Intelligent Design. HarperOne, 2013. Este libro aborda en profundidad la falta de formas intermedias en el registro fósil del Cámbrico y presenta el fenómeno como un reto significativo para el modelo de evolución gradual.

Gould, S. J. Wonderful Life: The Burgess Shale and the Nature of History. W. W. Norton & Company, 1989. Gould, un reconocido paleontólogo, describe la "Explosión del Cámbrico" y reflexiona sobre la complejidad de los organismos encontrados en ese periodo, reconociendo la dificultad que esto presenta para el gradualismo darwiniano.

Conway Morris, S. The Cambrian "Explosion": Slow-fuse or megatonnage? Proceedings of the National Academy of Sciences, vol. 97, no. 9, 2000, pp. 4426-4429. Este artículo científico analiza el evento del Cámbrico y sugiere que la aparición súbita de especies complejas requiere reevaluar los mecanismos evolutivos propuestos.

Valentine, J. W., Erwin, D. H., & Sepkoski, J. J. The Cambrian Explosion: The Construction of Animal Biodiversity. American Scientist, vol. 85, no. 2, 1997, pp. 126-137. En este artículo, los autores examinan la "Explosión del Cámbrico" y cómo esta fase de biodiversificación desafía la narrativa evolutiva convencional.

These sources explore the impact of the "Cambrian Explosion" on the fossil record and the debate over its compatibility with the theory of gradual evolution.

65) Fossils in suffocation position:

Many fossils of animals, including dinosaurs, have been found in what appears to be a choking position, suggesting that they died suddenly and were quickly buried by sediment.

This is consistent with a catastrophic global flood event, such as the Flood described in the Bible, and not with slow sedimentation processes over millions of years.

It is also a classic in the debates on creationism and the universal flood. Many fossils of animals, including dinosaurs, have been found in positions that suggest asphyxiation, indicating that they died suddenly and were quickly buried by sediment. This type of preservation is consistent with a catastrophic event, such as a major flood, rather than a slow sedimentation process that would take millions of years.

The discovery of fossil remains supports the possibility that events such as the Flood described in the Bible may have been responsible for the mass death and rapid burial of these creatures, which contrasts with evolutionary interpretations that propose long periods of time for fossil formation.

Las fuentes que abordan el tema de los fósiles en posición de asfixia y su interpretación en el contexto de eventos catastróficos incluyen:

Behrensmeyer, A. K. Fossils in the Making: Vertebrate Taphonomy and Paleoecology. University of Chicago Press, 1982. This book discusses fossilization processes and how the positions and conditions of fossils can reveal clues about the death and burial of organisms, including sudden deaths and rapid burials.

Brand, L. R. Faith, Reason, and Earth History: A Paradigm of Earth and Biological Origins by Intelligent Design. Andrews University Press, 2009. Brand examines the fossil evidence and discusses how the position of some fossils may be consistent with a large-scale catastrophic event, such as the Flood.

McGowan, C. The Raptor and the Lamb: Predators and Prey in the Living World. Princeton University Press, 1997. This text explores fossils that suggest sudden deaths, including fossils of vertebrates in agony postures, and questions interpretations exclusively based on slow and gradual processes.

Snelling, A. A. Earth's Catastrophic Past: Geology, Creation, and the Flood. Institute for Creation Research, 2009. Snelling argues that many fossils found in a suffocating position support the possibility of rapid burial events, such as the biblical Flood.

These sources examine preservation patterns in fossils and how they might be interpreted as evidence of sudden death and burial events, as opposed to gradual fossilization processes.

66) Human Presence in Coal Layers.

The "human presence in coal layers" argument is based on finds of human artifacts, footprints, and other remains that are apparently embedded in coal strata, which, according to evolutionary chronology, would have formed millions of years ago, long before humans existed.

These types of discoveries are controversial, since, if authentic, they would defy the time scale of evolution by placing humans in a much older period than the conventional model suggests.

If these findings are confirmed as genuine, they would indicate that the traditional evolutionary chronology might have significant limitations or errors.

This apparent temporal coexistence suggests a possible view of Earth's history in which humans and these coal strata are not so far apart in time, which would be consistent with an interpretation that considers a recent history for humanity and Earth.

They can be consulted in works by scholars of alternative chronology, publications by institutions such as the *Institute for Creation Research (ICR)* and *Answers in Genesis*, where alternative interpretations and critical analyses of geological chronology and the dating of fossils and sediments are reviewed.

67) Recent Origins of Agriculture

It's hard to imagine that human beings, with intelligence comparable to our own, could have existed for tens or hundreds of thousands of years without discovering something as fundamental as growing plants from seed.

However, the archaeological record indicates that agriculture began less than 10,000 years ago, which poses a contradiction with evolutionary chronology, which places anatomically modern humans on Earth from approximately 200,000 years ago.

This fact is intriguing, since, if evolution were correct, it would seem unlikely that humans, having reached advanced levels of intellectual capacity, would have taken so many millennia to discover agriculture and the domestication of plants.

According to archaeological studies, agriculture appears remarkably about 10,000 years ago in the Fertile Crescent, supporting the idea of a recent development of civilization and suggesting that evolutionary times might be overestimated.

The *Fertile Crescent* is a historical region in the Middle East that encompasses areas of present-day Iraq, Syria, Lebanon, Israel, Palestine, Jordan, and northeastern Egypt, especially in the areas near the Tigris and Euphrates rivers, as well as in the Nile Valley.

This region is known for being one of the first areas where agriculture and the domestication of plants and animals emerged approximately 10,000 years ago, marking the beginning of agricultural civilization and the transition to more settled and organized societies.

From the perspective of biblical chronology, the beginning of agriculture about 10,000 years ago in the Fertile Crescent corroborates and confirms the Genesis accounts, which describe the work of the land following the expulsion of Adam and Eve from Eden.

The development of agriculture and the organization of early agricultural societies in this region relates to the biblical story of early humans cultivating the land as part of their post-Edenic life.

This view suggests that the emergence of agricultural practices supports a model of humanity that has a relatively recent origin, in accordance with the biblical narrative of creation, and reinforces the idea of a humanity designed for the purpose of dominating and caring for the earth, as mentioned in *Genesis*.

Suggested sources:

Diamond, J. (1997). *Guns, Germs, and Steel: The Fates of Human Societies*. W. W. Norton & Company. This book analyzes the expansion of agriculture and its recent origins from a historical perspective.

Ochoa, G. (2011). *The Origins of Agriculture: An International Perspective*. Cambridge University Press.

68) Mutations and Natural Selection

Evolutionists should recognize that mutations are the only source of new genetic information for natural selection to operate.

According to the ESPASA dictionary, a mutation is "the alteration produced in the structure or number of genes or chromosomes of an organism, which is transmitted by inheritance".

Dr. H.J. Müller, Nobel laureate for his work on mutations, noted: "Exhaustive examinations show that the vast majority of mutations are detrimental to the organism in its task of surviving and reproducing... THE GOOD ONES ARE SO RARE THAT WE CAN CONSIDER THEM ALL AS BAD" (Bulletin of Atomic Scientists, 11:331).

This means that beneficial mutations, those that would be useful for evolution, are extremely rare.

It is important to note that a mutation only passes on to future generations if it occurs in the reproductive cells (sperm or eggs).

In addition, the probability of even one sequence of 5 beneficial mutations occurring in the same cell is so low (1 in 100 quadrillion) that, even in a population of 100 million organisms with daily reproductive cycles, this event would only happen once every 274 billion years.

In the face of these extremely low probabilities or "remote odds," it takes more faith to believe in evolution through mutations than to accept the existence of a creator.

H.J. Müller, winner of the Nobel Prize in Physiology or Medicine in 1946, made outstanding contributions to the study of mutations and the effects of radiation on genes. His observation about the rarity of beneficial mutations comes from his research, and the quote used is in the Bulletin of the Atomic Scientists (11:331). Their work explores how the vast majority of mutations tend to be harmful to the body.

Bulletin of the Atomic Scientists, which presents the perspective of Müller and other scientists in relation to the effects of mutations. This newsletter provides analysis on the impact of mutations and has been a source of scientific debate on genetic stability in contexts of environmental change and evolutionary adaptations.

Population genetics and studies such as those of Motoo Kimura, who introduced the neutralist theory of molecular evolution, proposing that most mutations are neutral or harmful. Their work explores how gradual evolution based on beneficial mutations is statistically unlikely and cannot explain genetic diversity on its own.

"Molecular Biology and Evolution" and other journals of molecular biology and genetics regularly publish studies on the frequency and impact of mutations, showing that genetic changes necessary for evolution are unlikely and that beneficial mutations are extremely rare (see, for example, Molecular Biology and Evolution, articles on mutations and their impact on adaptation).

John Sanford in his book Genetic Entropy explores how the accumulation of harmful mutations challenges a population's ability to evolve into more complex structures, arguing that the number of harmful mutations far outweighs that of beneficial mutations.

These sources help to consolidate the argument that the evolutionary process based solely on mutations is statistically improbable and requires a reconsideration of the proposed mechanisms for evolution.

69) Fossils and Rapid Formation.

To support this argument, geological studies have shown that fossilization requires specific conditions, mainly the rapid covering of the organism by sediments, which prevents its decomposition. Without this immediate coverage, the organism disintegrates due to factors such as predator action, decomposition, and bacterial activity.

A relevant example of rapidly formed fossils is seen in the eruption of Mount St. Helens in 1980, where layers of sediment were rapidly deposited and contained fossil remains in just days or weeks. This event showed how catastrophic conditions can create geological strata in a short time, which supports the possibility of sudden geological events in the past.

In addition, polystratus trees have been found in the Joggins region of Nova Scotia that pass through multiple layers of sediment, reinforcing the idea that these strata may have formed in rapid succession. If each layer represented millions of years, the trees would have decomposed before being completely covered. Instead, the arrangement of these fossils suggests rapid sediment deposition, more consistent with catastrophic events such as floods. This type of formation is difficult to explain under the evolutionary model of slow and gradual sedimentation, which supports a shorter chronology and points to events of rapid geological transformation.

This pattern is more consistent with a catastrophic event, such as the global Flood, rather than the slow, gradual processes that postulate millions of years for the formation of geological strata.

Brand, L. R., & Florence, M. L. (2000). "The Fossil Record and Flood Geology." Origins Journal.

Morris, J. D. (2010). The Young Earth.

Gibling, M. R. & Rygel, M. C. (2008). "Joggins Fossil Cliffs: Nova Scotia's Carboniferous Park." Atlantic Geology Journal.

Coffin, H. G. (1983). "Origins of the Joggins Polystrate Fossils." Creation Research Society Quarterly.

70) On the same amber.

Argument: Amber, formed from the resin of trees, is a substance that results from a relatively rapid process of hardening and polymerization.

Initially, resin is a viscous substance, which hardens quickly to become amber under optimal conditions, when it is buried and without contact with oxygen, thus preventing its decomposition.

This natural preservation process raises questions about amber's ability to remain intact for millions of years, since, over such long times, variations in pressure, temperature, and other environmental factors could affect its structure and molecular stability.

Over millions of years, amber should be subject to internal degradation processes, as well as chemical reactions with surrounding minerals.

The fact that amber remains unaltered, preserving even its translucency and color, is more consistent with a more recent chronology, suggesting that these deposits are younger than is usually claimed in traditional geological time models.

For the argument about the preservation of amber and its chronology, one can consult studies in the chemistry of organic materials and publications on the formation of amber in specific geological environments. Researchers such as Andrew C. Scott and his team in the article "Fossil Resins: An Overview," published in *Geology Today* (2004), have addressed the formation and preservation of fossil resins and their rapid transformation into amber under specific conditions. These studies highlight the stability of the resin and its preservation under quick-burial conditions, which can be interpreted in the context of a more recent chronology.

For additional details on the processes that enable the preservation of amber, the work of Robert N. Clarke, in "Amber: Structure and Chemistry of the Fossilized Resin," in the *Journal of the Geological Society* (1993), also provides technical information on the rapid transformation of resin into amber.

71) Marine Fossils in the Mountains.

Plot: The discovery of marine fossils on mountaintops, such as the Andes, provides clear evidence that these areas were submerged underwater in the past. A striking example is the discovery of 500 giant oysters in the Peruvian Andes, at almost 4,000 meters above sea level. These oysters, some up to 3.5 meters in circumference and weighing approximately 300 kilograms, show that the area was covered by water at some point.

Another striking case is Mount Everest, where fossils of sea creatures have been found in the rocks of its summit, suggesting that even the world's tallest mountains were once under the sea.

The existence of marine fossils in mountains and mountain ranges reinforces the idea of a catastrophic event such as the global Flood, capable of flooding even the highest areas.

This scenario is difficult to explain under the evolutionary assumptions of millions of years of slow tectonic uplifts, but it is consistent with a massive flood that rapidly covered the Earth, as described in the Bible.

Plot: The order in which fossils are found in the strata corresponds to their natural habitats, reinforcing the idea that the strata formed rapidly during an event like the Flood and not over millions of years.

The argument is based on the observation that fossils in geological strata do not follow a gradual evolutionary pattern but are arranged according to the habitats where the creatures lived.

For example, fossils of marine animals, such as mollusks and fish, are found in the deeper layers, while fossils of land animals are found in the upper layers. This order is most consistent with a rapid accumulation of sediment during a catastrophic event, such as the Flood, which would have buried ecosystems in sequence, from the seafloor to the higher ground.

A famous experiment conducted by Gilbert Hall showed that sediments and aquatic creatures tend to be arranged according to their density and natural location when swept away by water currents. This kind of rapid deposition fits better with the narrative of a global flood than with millions of years of slow, gradual deposition. Therefore, the order of fossils by habitat reinforces the idea that geological strata were formed in a short time.

Finds of marine fossils in mountainous areas, such as Everest and the Andes, are documented in the work of *Earle E. Spamer* and *Raymond G. Bernor* in "Historical Biogeography of High-Altitude Marine Fossils," in *Bulletin of the American Museum of Natural History* (1990). In addition, Gilbert Hall has conducted experiments on the arrangement of fossils in aquatic sediments, which are discussed in sedimentology studies and catastrophic events such as the Flood, available in publications of *the Creation Research Society Quarterly*.

Part VI: Chemical Evidence

72) Anomalies in Radioactive Decay of Uranium

Uranium, particularly the isotope uranium-238, is widely used in radiometric dating methods due to its long half-life.

However, several studies have revealed possible fluctuations in the decay rate of this element, calling into question the reliability of dating methods that assume that this rate is constant over millions of years.

Under laboratory conditions, researchers have observed that factors such as variations in the Earth's magnetic field, exposure to intense radiation, and changes in pressure could affect the decay rate of uranium-238.

These variations raise doubts about the assumption of absolute stability in the radioactive decay of uranium throughout geological history.

Experimental studies by scientists in the 2000s suggest that the decay rate of uranium may not be immutable.

This implies that, under certain geological or cosmic conditions, the decay rate of uranium could have been altered, leading to possible errors in age estimates based on this isotope.

This situation poses a fundamental limitation to dating models that assume that radioactive decay has remained constant for billions of years.

Proponents of a young Earth use these findings to question the validity of dating techniques and argue that calculations based on uranium decay could have overestimated the age of certain rocks and fossils.

Fischbach, E., Jenkins, J. H., & Sturrock, P. A. (2009). *Evidence for Correlations Between Nuclear Decay Rates and Earth-Sun Distance. Astroparticle Physics, 31(6), 407–411.*

Emery, G. T. (1972). *Perturbation of Nuclear Decay Rates. Annual Review of Nuclear Science, 22(1), 165–202.*

73) Radioactive Decay Abnormalities of Thorium

Argument: Thorium, and specifically the isotope thorium-232, is also used in radiometric dating techniques.

This isotope has a long half-life, but, like uranium, recent research has revealed that the decay rate of thorium may not be absolutely constant.

Studies have shown that thorium can be sensitive to extreme pressure variations, elevated temperatures, and the presence of certain minerals.

These conditions, although unusual, could affect the stability of thorium and modify its decay rate, which calls into question the accuracy of dating techniques based on this element.

In addition, experimental observations have indicated that thorium-232, under certain environmental conditions, could chemically react with surrounding materials, thus altering the dating results.

This anomaly poses a limitation to radiometric dating techniques that assume the immutability of the decay rate of thorium over millions of years.

As in the case of uranium, these variations in thorium's decay rate reinforce the stance of a more recent chronology for Earth, suggesting that the ages assigned to certain rocks and minerals could be overestimated.

O'Brien, K. (2008). Variability in Nuclear Decay Rates: An Overview and Implications for Geochronology. Journal of Environmental Radioactivity, 99(1), 127–132.

Becker, H., & Clayton, R. N. (2000). Th/U Dating and Isotope Ratios in Zircons: Implications for Radioisotope Decay Constants. Earth and Planetary Science Letters, 177(3-4), 453–469.

74) Rapid sedimentation in fossils:

Argument: Rapid sedimentation in fossils suggests that fossils formed over a short period of time, rather than over millions of years as evolutionary theory proposes.

In order for an organism to fossilize, it needs to be quickly buried under layers of sediment, thus protecting it from decay and predators.

This is observed in large fossil burial grounds, where numerous organisms are preserved almost perfectly, indicating a rapid and massive burial process.

Catastrophic events (such as the flood, in fact this is proof of it, as it is observed everywhere) but such as floods, volcanic eruptions or tsunamis, are plausible scenarios that would have allowed this accelerated preservation.

Many fossils, found in "asphyxiation" positions or in large groups with no signs of decay, support the idea of a quick burial.

In cases of slow fossilization processes, the remains would have been exposed to factors that accelerate decomposition and fragmentation before being buried.

Sudden fossilization, however, points to events of great magnitude, such as a flood, capable of dragging and burying organisms in a matter of hours or days.

This type of fossilization is consistent with a shorter timeline for the formation of geological layers, thus defying the proposed times of millions of years.

Brand, L. R., & Tang, T. (1991). Fossil Vertebrates and Their Taphonomy in Lacustrine and Fluvial Deposits, Lake Gosiute (Eocene), Wyoming. Palaeogeography, Palaeoclimatology, Palaeoecology, 81(1-2), 81-96.

Fastovsky, D. E., & Sheehan, P. M. (2005). The Extinction of the Dinosaurs in North America. GSA Today, 15(3), 4-10.

Part VII: Cosmological, Stellar, and Design Evidence

75) Recent Supernovae:

The lack of supernova remnants in the universe suggests that it could be much younger than currently estimated.

Supernovae, which are massive stellar explosions at the end of the life of certain stars, hurl large amounts of matter and energy into space, leaving a visible remnant.

If the universe really were billions of years old, as is believed, we should find a large number of these supernova remnants distributed in space.

However, the number of observed remnants is surprisingly low.

Current theory indicates that supernovae should occur in galaxies like ours about once every 50 years, leaving behind debris that would last thousands or millions of years.

The scarcity of these remnants has led some researchers to suggest that the universe could be considerably younger than evolutionary theory proposes, which would explain the lack of evidence regarding the expected abundance of these remains in space.

Davies, P. C. W., & Lineweaver, C. H. (2004). The Arrow of Time and the Nature of Space-Time. Studies in History and Philosophy of Science Part B, 35(1), 3-19.

Clark, D. H., Caswell, J. L., & Green, A. J. (1976). A Study of Supernova Remnants in the Galaxy. Monthly Notices of the Royal Astronomical Society, 175(1), 1-19.

76) Accelerated Stellar Evolution.

The rapid stellar transformations observed in the universe defy the long times proposed by evolutionary theory.

Accelerated stellar evolution refers to documented cases in which stars have undergone major changes in much shorter time spans than traditionally estimated.

According to evolutionary theory, stars should develop extremely slow phases of life spanning millions or even billions of years. However, recent observations reveal transformations in some stars that occur in significantly shorter times.

For example, certain types of variable stars have shown drastic changes in brightness and size over decades, which is incompatible with conventional stellar models.

These observations challenge the assumptions of conventional stellar theory, suggesting that stars may evolve at faster rates than expected.

This possibility is consistent with a much younger universe than is proposed in evolutionary theory.

Schaefer, B. E., & Pagnotta, A. (2012). "An absence of ex-companion stars in the type Ia supernova remnant SNR 0509-67.5." Nature, 481(7380), 164-166.

Meynet, G., & Maeder, A. (2005). "Stellar evolution with rotation." Annual Review of Astronomy and Astrophysics, 43, 581-634.

77) Venus and its Mountains.

Venus, despite its proximity to the Sun and extreme temperatures, has mountains that shouldn't exist if the planet were billions of years old, as the crust would have softened. This supports the idea of a young planet.

Venus is a planet close to the Sun, with extremely high temperatures reaching more than 460 degrees Celsius.

These temperatures should have caused the planet's crust to soften considerably if Venus were billions of years old, as the conventional theory of the age of the solar system proposes. However, the mountains and other geological features on Venus' surface present a challenge to this idea.

These mountains, some of which are as large as Earth's, shouldn't have remained stable for so long if Venus's crust had been subjected to such high temperatures for billions of years.

The extreme conditions should have caused the surface to flatten or warp. The existence of these geological formations in their current state suggests that Venus has not had enough time for these warping processes to occur, which supports the idea of a much younger planet than previously thought.

This argument reinforces the view of a young solar system, where celestial bodies like Venus have not had the necessary timescales for the geological processes proposed by evolutionary theories to be completed, suggesting a more recent chronology.

Basilevsky, A. T., & Head, J. W. (2003). "The Geologic History of Venus: A Stratigraphic View." Journal of Geophysical Research: Planets, 108(E6). This study discusses the geological features of Venus, including its surface and mountainous structures, in the context of planetary geology and high-temperature conditions on Venus.

Schaber, G. G., et al. (1992). "Geology and Distribution of Impact Craters on Venus: What are they Telling Us?" Journal of Geophysical Research: Planets, 97(E8), 13257-13301. This paper reviews the surface structure of Venus, highlighting the unexpected stability of its geological formations under high-temperature conditions, and discusses the challenges this poses to theories of long-lived planetary evolution.

78) Balance of the Solar System.

The precise location of the Earth in relation to the Sun and the Moon, as well as the balance in the rotation of our planet, allow life.

A slight change in any of these factors would make life impossible. This perfect balance suggests intentional design and recent creation.

Earth is in what scientists call the "habitable zone" or "Goldilocks zone," a region in space around a star where conditions are just right to allow life.

This balance is crucial for the existence of liquid water, moderate temperatures, and a stable atmosphere, which are essential conditions for life as we know it.

The exact distance between the Earth and the Sun is critical. If we were a little closer to the Sun, the Earth would be too hot, and the water would evaporate, creating an uncontrolled greenhouse effect similar to that of Venus.

Conversely, if we were farther away, temperatures would be too cold, and the water would freeze, making life impossible. In addition, the Moon plays a key role in stabilizing the tilt of the Earth's axis, which in turn regulates the climate and seasons, conditions that make the development of complex ecosystems possible.

Balance in the Earth's rotation is also crucial. Our rotation speed is perfect for avoiding weather extremes; If we rotated faster, the winds would be extremely strong, and if we rotated slower, the days and nights would be too long, making it difficult to regulate temperatures.

These factors, in addition to the perfectly tuned gravity between the Sun, Earth, and Moon, show a precise fit that allows for life.

A small change in any of these parameters would result in an inhospitable planet. This exact balance suggests an intentional design and not a result of random processes over billions of years.

From this perspective, it is proposed that the solar system is much younger than evolutionary models suggest and that it was designed to sustain life from the beginning.

Gonzalez, G., & Richards, J. W. (2004). The Privileged Planet: How Our Place in the Cosmos is Designed for Discovery. Regnery Publishing. This book explores the notion of the "habitable zone" and the fine-tuning of the factors that enable life on Earth, analyzing the balance in the solar system and the location of the Earth.

Ward, P. D., & Brownlee, D. (2000). Rare Earth: Why Complex Life is Uncommon in the Universe. Copernicus. This book discusses Earth's exceptional life-sustaining conditions and the improbability that these features arose randomly, suggesting the possibility of intentional designl.

Laskar, J., & Robutel, P. (1993). "The Chaotic Obliquity of the Planets." Nature, 361(6413), 608-612. This study reviews the stabilizing role of the Moon in the tilt of the Earth and how this affects climate and seasons, emphasizing the precision needed in the Earth-Moon system to allow life.

79) Expansion Speed of the Universe.

The expansion rate of the universe is exact to allow the formation of galaxies and planets. A faster or slower expansion would have made the existence of solar systems impossible. This fine-tuning is evidence of a young, designed universe.

The universe is expanding at a precise rate, a phenomenon known as the Hubble constant.

This rate of expansion is critical for the formation and stability of galaxies, solar systems, and ultimately life.

If the universe had expanded faster from the start, matter would not have clumped together to form stars and galaxies, and space would be a vacuum with no stable solar systems.

On the other hand, if the expansion had been slower, gravity would have caused the universe to collapse in on itself long before stars and planets could form.

This delicate balance is known as the "fine-tuning" of the universe. The forces that govern the expansion of the cosmos, such as gravity and dark energy, are tuned in such a way as to allow complex structures to exist.

The probability that these factors were adjusted randomly is extremely low, which leads many to think that this adjustment is not the result of chance, but points to an intentional design.

The fact that the universe has reached a state where it can host galaxies, stars, and planets like Earth suggests a purpose in its creation. Some argue that this fine-tuning is evidence that the universe is not as old as conventional models suggest but was created precisely and recently to allow life as we know it.

Rees, M. (2001). Just Six Numbers: The Deep Forces that Shape the Universe. Basic Books. In this book, physicist Martin Rees explores the "fine-tuning" of universal constants, such as the Hubble constant and its influence on the stability of cosmic structures, highlighting the precision required in the expansion of the universe to allow the formation of galaxies and star systems.

Tegmark, M., & Rees, M. (1998). "Why is the Cosmic Microwave Background Fluctuation Level 10^-5?" The Astrophysical Journal, 499(2), 526-532. This paper explains the necessary adjustment in the expansion rate of the universe for the formation of galaxies and structures, and how a minimal variation in this constant would have significantly altered the development of star systems.

Barrow, J. D., & Tipler, F. J. (1986). The Anthropic Cosmological Principle. Oxford University Press. This classic text addresses the fine-tuning of the physical constants and the rate of expansion of the universe, arguing that these conditions are compatible with the existence of stable life and structures in the cosmos.

80) CFCs in Polar Ice

Argument: The presence of chlorofluorocarbons (CFCs), compounds created and released into the environment since the twentieth century, has been identified in polar ice cores.

These CFCs are trapped in ice sheets, indicating that these sheets form and accumulate more rapidly than suggested by chronologies that assign thousands of years to ice sheets in Antarctica and Greenland.

If these ice sheets were indeed the product of slow accumulations over hundreds of thousands of years, traces of recent CFCs would be expected to be more superficially distributed.

However, their presence in deeper layers implies a much faster rate of ice accumulation, which supports the idea of a more recent chronology for these ice formations.

Jaworowski, Z. (1994). "Ancient Atmosphere—Validity of Ice Core Records." Environmental Science & Pollution Research, 1(3), 161-171. This paper examines the interpretation of ice cores and questions the accuracy of extended chronologies, indicating that modern pollutants, such as CFCs, are found at deeper levels than expected.

Taylor, K. C., Alley, R. B., & Grootes, P. M. (1992). "Low-resolution timescales and the age of the Taylor Dome Ice Core." Geophysical Research Letters, 19(19), 1991-1994. In this study, they analyzed the depth of certain pollutants, such as CFCs, in the Taylor Dome ice, suggesting that the chronologies of thousands of years could be re-evaluated based on the current rate of accumulation.

81) Changes in Sea Level and Coastal Strata

Geological research has shown that changes in sea level can cause rapid strata to form in coastal areas.

These strata, often composed of marine sediments and entrained materials, accumulate in much shorter times than is commonly considered in evolutionary chronologies of millions of years.

During sea-level change events, such as rapid rises, well-defined strata have been observed, supporting the hypothesis that sedimentary layers can form in short times, even decades or less, thus challenging the idea of slow deposition.

This phenomenon is clearly visible in coastal areas and deltas, where intense storms, tsunamis, or rapid sea-level rises have deposited large amounts of sediment at documented times.

A prominent example is the 2004 tsunami in the Indian Ocean, which deposited layers of sand and sediment up to 30 cm in some coastal areas in minutes, creating visible strata without the need for long periods of time.

Hurricane Katrina in 2005, which formed sedimentary deposits in the Louisiana region and the Mississippi River Delta, where rapid changes in water and sediment generated strata over large areas in a matter of days.

In 2011's Japami left deposits more than 40 cm thick in certain coastal areas of Sendai and Tōhoku, demonstrating how a single event can generate defined and distinctive layered strata in minimal time.

These examples of massive sedimentary in response to extreme events support the possibility that similar formations in the geological past, today observed as strata, could have formed rapidly, thus supporting a more recent chronology against the assumption of gradual processes of millions of years and consistent with the Flood.

Don J. Easterbrook, "Surface Processes and Landforms", Posted by Prentice Hall, it analyzes how changes in sea level influence the formation of sedimentary strata on coasts and continental shelves.

J.R.L. Allen, "Principles of Physical Sedimentology", Champman & Hall, it examines the conditions and times when sediments accumulate during rapid changes in sea level, which supports the rapid formation of strata in these environments.

C. Vita-Finzi, "The Mediterranean Valleys: Geological Changes in Historical Times", analyzes how changes in Mediterranean levels generated strata in recent historical times.

This evidence reinforces the possibility of rapid geological formation and suggests that many sedimentary strata could have originated in much shorter times than traditional evolutionary chronologies assume.

Conclusion

Throughout this book, "81 Evidence Pointing to an Earth and a Young Universe: Refutation of Evolutionary Time," we have pored over the scientific evidence (not supported by biblical quotes) that challenges the prevailing narrative of a 4.6-billion-year-old Earth. Although these do support them.

From astronomical records to geological and biological evidence, each piece of evidence presented offers a distinct but complementary perspective that supports the idea of a young world, created by design, rather than the result of slow, accidental natural processes.

The reader will find solid arguments based on direct observation and scientific studies that do not always receive the same visibility.

This work is not only a call to question the paradigm established by an apparent dictatorship in the information we receive from Basic, but also an invitation to explore the possibility that Scripture and science can be harmonized in a more coherent way.

This search for truth transcends the academic and touches the spiritual, offering answers to both believers and those seeking greater understanding about the origin of our world.

Each of the 81 pieces of evidence forcefully refutes the idea of a billion-year-old Earth and presents a firm defense of the belief in a young creation.

In the end, science is not only a tool to discover the world, but an opportunity to know its Creator better. Perhaps you could refute some of the evidence mentioned here, but it is impossible to refute all of it. And one of them, in isolation, is enough to have us already doubted.

Those who defend evolution and the extended time of billions of years often resort to complex and voluminous explanations to justify their position, but what is simple and clear does not require embellishments. God, in His sovereignty, expressed it in a single sentence: "In the beginning God created the heavens and the earth" (Genesis 1:1). There are no ambiguities, his Word is direct. While those who defend chance and prolonged time need millions of words and studies to sustain their belief in a blind and random process.

What is in sight, as the adage goes, needs no explanation. The complexity and order of the universe, life itself, and the design so evident in all that surrounds us are testament to an intelligent mind behind creation. There is no need to develop complicated theories when the truth is right before our eyes.

Of course, despite the obvious, there will always be those who try to think about the matter, looking for ways to explain what they cannot, because the divine truth confronts their beliefs. But at the end of the day, it's about believing or not believing, accepting what God has revealed, or clinging to human theories that are constantly changing and adjusting to try to square with what science can't fully explain.

The simplicity of truth does not make it any less profound or valuable. On the contrary, it reflects the magnificence of a God who does not need complicated embellishments to express the truth, for His work is perfect and available to those who seek it with an open heart.

The "Ockham's razor" (or "Occam"). It was proposed by the English friar and philosopher William of Ockham (William of Ockham) in the fourteenth century. This principle states that, among several possible explanations for a phenomenon, the simplest tends to be the correct one. It is not an absolute postulate that guarantees truth, but it is a useful guide in the search for rational explanations and is widely used in science, philosophy, and logic.

I abide by this principle, because it takes more faith to believe the far-fetched, sophisticated, complicated, and hardly credible attempts to find an explanation for what was created on an intricate soup or azopao of chance and time, than the miraculous simplicity of the Bible.

The points I mention reflect a process of searching for and organizing information over the years and highlight the idea that many discoveries have come to me by chance, something known as "serendipity" (the act of finding something valuable and interesting unexpectedly, by chance or accident).

This serendipity, combined with my immersion and openness to the world of creationism, allowed me to collect a number of pieces of evidence that defy evolutionary chronology. But it's not that I deserve a Nobel Prize or that I want Clair Cameron Patterson's 1956 prize. I haven't discovered anything. I have organized accumulated information only.

It is interesting how, being in different "worlds" (be it creationism or evolutionism), people tend to find only the evidence that reinforces their own beliefs, leading to a biased or incomplete view. This "dictatorship of information," as I call it, filters what people see, keeping them in a state of indoctrination that prevents objective discussion. In my case, after 30 years of Christian walking, and perhaps because of the time of rest after a recent surgery, I found an opportunity to organize everything I had accumulated into a well-founded compendium.

For me, the 9 most outstanding pieces of evidence among the 81 that I have gathered, there are several that are especially intriguing because of their solidity and how difficult it is to refute them. Some of these could include:

1. Short-lived comets: Comets lose mass rapidly whenever they approach the Sun. If the solar system were billions of years old, these comets should have disappeared long ago. However, its current existence defies that timescale.

2. Moon's retreat: The Moon moves away from the Earth about 3.8 cm per year. If this process had been going on for billions of years, the Moon would have been too close to Earth in the past, leading to unsustainable gravitational problems.

3. Dust on the Moon: The dust layer on the Moon is much smaller than expected if we assume that the Moon is billions of years old. This suggests that the Moon, and by extension the solar system, is much younger.

4. Shrinkage of the Sun: Some studies suggest that the Sun is losing mass and shrinking in size. If this process had occurred over billions of years, the Sun would have been too large in the past, which would have prevented life on Earth.

5. Salinity of the oceans: The salinity of the sea increases as rivers and other processes carry salt into the ocean. If this process had been going on for billions of years, the oceans should be much saltier than they are today.

6. Erosion of the continents: The current rate of erosion should have disintegrated the continents several times if the Earth were billions of years old. However, the continents still remain, suggesting a younger age.

Helium   in the atmosphere: Helium escapes into space over time. The amount of helium in the atmosphere doesn't match what we'd expect if Earth were billions of years old.

8. DNA evidence in ancient fossils: The presence of DNA and proteins in supposedly ancient fossils poses a challenge, as these biomolecules break down quickly. Their existence in fossils thought to be millions of years old suggests that those fossils could be much younger.

The   Cambrian Explosion: During the Cambrian period, many complex species appeared abruptly in the fossil record with no clear evolutionary precursors. This "explosion" of complex life is difficult to explain under models of gradual evolution, suggesting a more recent creation event.

One of the most critical and controversial points in the debate over the age of the Earth has to do with the assumptions on which dating methods using radioactive isotopes are based.

These methods, such as carbon-14 dating or uranium-lead dating, depend on certain fundamental premises that, if not met, could invalidate the results obtained.

Another fascinating example that calls into question the chronology of millions of years is the case of fossilized trees, called "polystratified trees," that traverse several geological strata that, according to conventional dating, would have formed in different eras separated by thousands or millions of years.

If these strata really represented such long times, how is it possible that a single tree is present throughout so many layers? A tree cannot remain upright for thousands of years without decomposing before being fully fossilized.

This phenomenon suggests that the strata could have been deposited quickly, as a result of a catastrophic event, such as a massive flood (flood) or a series of volcanic eruptions, which would support the idea of a younger Earth than commonly thought.

Not only are these points difficult to refute from an evolutionary perspective, but they also open the door to the possibility that the Earth and the universe are much younger than conventionally believed. This compendium of evidence, based on these and other observations, challenges the prevailing narrative and offers a valuable perspective for those seeking to see the other side of the debate.

Es importante destacar que este tipo de trabajo no busca imponer una verdad absoluta, sino ofrecer un punto de vista que ha sido, en muchos casos, censurado o marginado en el discurso científico.

By making this evidence visible, the reader is invited to question the assumptions on which traditional knowledge is built, opening the door to new interpretations that, although they do not fit into the conventional paradigm, offer legitimate arguments that must be considered.

Many aspects of the evolutionary hypothesis about mass extinction, including the effects of the Chicxulub impact, for example, are based on estimates and assumptions.

Science depends on reconstructing past events from available evidence, but these inferences do not always offer absolute certainty.

Theories about the extinction event, therefore, are not a direct observation, but an interpretation based on current indications.

This type of scientific model is based on assumptions and, to some extent, on what some of us would call "faith" in the inferential method and in the consistency of the data collected.

The Author.

Explanatory Note on Sources:

It is common for many texts to include extensive lists of sources or bibliography at the end to support the above.

However, that approach can be tedious and impractical.

For this reason, I have opted for a more accessible structure, presenting the references at the end of each piece of evidence to allow immediate and precise consultation of each argument.

In addition, each approach, evidence or argument presented is, for the most part, independent of the others, which facilitates the particular consultation of each topic.

This book is the result of an extensive review of scientific literature, texts that I have studied and keep, and a collection of information and studies that I have collected on my computer over the years.

In addition, it includes relevant academic research and topics that came into my hands by chance and that, due to their interest, I decided to save, analyze and corroborate in more detail.

This work seeks to offer a critical view of the evidence that points to a young Earth, facilitating access to key references.

Unlike conventional academic analysis, this text is not intended to replace the original sources, but rather to guide those interested towards an understanding of the data and arguments presented.

For those who wish to dig deeper, I recommend reviewing the quotes included in each piece of evidence, and exploring more publications in the fields of geology, cosmology, and biology.

Much of the information is supported by an interpretative analysis of scientific data and by sources that are often cited in the debate between science and faith.

I invite readers to consult the references throughout the text and to conduct their own research on the relevant scientific studies.

This work is both a compilation of decades of information and the result of what I consider providential: a recovery time after surgery, which I call "God's time," was key to organizing this accumulated information and shaping it. Without this providential pause, this compendium may never have come to fruition.

In this sense, the difficulty of establishing a conventional bibliography is not a shortcoming, but a natural consequence of this approach, where the research process has been more a continuous learning process than a purely academic search.

The Author.

www.ingramcontent.com/pod-product-compliance
Lightning Source LLC
Chambersburg PA
CBHW020637220526
45464CB00001B/186